Jian Guo Zhou

Lattice Boltzmann Methods for Shallow Water Flows

Springer
Berlin
Heidelberg
New York
Hong Kong
London
Milan
Paris
Tokyo

Jian Guo Zhou

Lattice Boltzmann Methods for Shallow Water Flows

with 50 Figures

Springer

DR. JIAN GUO ZHOU
7 Cherry Lane
SALE
M33 4NF
United Kingdom

Email: jgzhou77@yahoo.co.uk

ISBN 978-3-642-07393-9

Cataloging-in-Publication Data applied for

Bibliographic information published by Die Deutsche Bibliothek
Die Deutsche Bibliothek lists this publication in the Deutsche Nationalbibliografie; detailed bibliographic data is available in the Internet at <http://dnb.ddb.de>.

Springer-Verlag Berlin Heidelberg New York
a member of BertelsmannSpringer Science+Business Media GmbH

http://www.springer.de

Camera ready by authors
Cover design: E. Kirchner, Heidelberg
Printed on acid-free paper 32/3141/as 5 4 3 2 1 0

献 给

妻子：李协平 （中国重庆奉节）
儿子：周煜然 (Felix Y. Zhou)
女儿：周理佳 (Clare L. Zhou)
父亲：周开武 （中国重庆奉节）

纪念我的母亲：寿明琼

Preface

The lattice Boltzmann method (LBM) is a modern numerical technique, very efficient, flexible to simulate different flows within complex/varying geometries. It is evolved from the lattice gas automata (LGA) in order to overcome the difficulties with the LGA. The core equation in the LBM turns out to be a special discrete form of the continuum Boltzmann equation, leading it to be self-explanatory in statistical physics. The method describes the microscopic picture of particles movement in an extremely simplified way, and on the macroscopic level it gives a correct average description of a fluid. The averaged particle velocities behave in time and space just as the flow velocities in a physical fluid, showing a direct link between discrete microscopic and continuum macroscopic phenomena.

In contrast to the traditional computational fluid dynamics (CFD) based on a direct solution of flow equations, the lattice Boltzmann method provides an indirect way for solution of the flow equations. The method is characterized by simple calculation, parallel process and easy implementation of boundary conditions. It is these features that make the lattice Boltzmann method a very promising computational method in different areas. In recent years, it receives extensive attentions and becomes a very potential research area in computational fluid dynamics. However, most published books are limited to the lattice Boltzmann methods for the Navier-Stokes equations.

On the other hand, shallow water flows exist in many practical situations such as tidal flows, waves, open channel flows and dam-break flows. The basic feature of the flows is that the vertical effect can be neglected compared with the horizontal one with a good approximation. This allows a considerable simplification in the mathematical formulation by replacing the vertical momentum equation with the hydrostatic pressure distribution. As a result, such flows are usually described with the shallow water equations. A numerical solution of the shallow water equations turns out to be a very successful tool in studying a wide range of flow problems occurring in ocean, environmental and hydraulic engineering, for instance, tidal flows in estuary and coastal regions, river, reservoir and open channel flows. In literature, there are many computational methods available for solutions of the shallow water equations such as finite difference method, finite volume method, finite element method and Godunov-type method. Usually, a special treatment is required in these numerical procedures for either convective term, depth computations or source terms. All of these methods are developed on the basis of direct solutions to the shallow water equations.

Since the lattice Boltzmann method is a modern numerical technique, it is necessary and natural to investigate how to use the method for solving the shallow water equations. The author studied the problem and developed a well-defined lattice Boltzmann model for shallow water flows with or without flow turbulence, namely LABSWE and LABSWE^{TM}. It follows out that the lattice Boltzmann method is simple, efficient and accurate for solution of the shallow water equations. Therefore, it is timely to write a book in order to introduce this elegant method into research field, educational area, engineering sector and consultancy organisations so that the method may be used to solve real life flow problems efficiently and accurately.

This book may be used as a research reference for scientist, a practical method for engineers and consultancy organisations, and a text book for both undergraduate and postgraduate students.

Peterborough, June 2003 *Jian Guo Zhou*

Contents

1

Introduction

1.1 Outline of the Book

The book addresses a modern numerical technique, the lattice Boltzmann method (LBM). It consists of seven chapters. In this chapter, a background is given. This includes the origin of the lattice Boltzmann method, its advantages and the purpose of the book. Chapter 2 describes the governing equations for general fluid flows and subgrid-scale stress model for turbulence modelling. Based on these, the shallow water equations are derived in detail. This is the mathematical model for shallow water flows. In Chapter 3, we provide the detailed description of the lattice Boltzmann model for the shallow water equations (LABSWE). The Chapman-Enskog procedure is used to prove that the physical variables generated from the LABSWE are a solution to the shallow water equations. In Chapter 4, a centred scheme is presented for accurate treatment of force terms in the lattice Boltzmann equation. The feature of the scheme is analyzed and compared with available schemes. Also, the *necessary property* ($\mathcal{N}$-property) as a basic criterion for a numerical scheme is introduced. Chapter 5 describes a lattice Boltzmann model for the shallow water equations with turbulence modelling (LABSWE^{TM}) by incorporating the subgrid-scale stress model into the lattice Boltzmann equation. In Chapter 6, various boundary and initial conditions are described. Chapter 7 presents a few of numerical results to illustrate the efficiency, accuracy and capability of the lattice Boltzmann methods for shallow water flows.

1.2 History

With the development of computer techniques, it is possible to formulate a simple model for complex systems. This is the basic idea behind a finite discrete space-time model, i.e. a cellular automata (CA), a lattice gas automata (LGA) and a lattice Boltzmann method. In particular, the lattice Boltzmann method is formulated as a simple, efficient and accurate model for fluid flows.

Although there are close relations among the three models, they are developed into three independent research areas. In this section, the brief history is addressed.

1.2.1 Cellular Automata

A cellular automata was originally proposed to simulate life by von Neumann in the late 1940's. Since then, it has been further developed and becomes a very powerful tool in simulating various scientific problems [1, 2]. It is a simple model which is characterized by a set of synchronized identical finite automata with the same evolution rules. The basis of the cellular automata is that space is divided into cells and time is discrete. Each of the cells has initial state and will be updated according to simple rules. For example, Conway [3] applied a cellular automata to create his famous *Game of Life*. He used 2D square lattice like a checkerboard and defined initial state for each cell as either alive with "1" or dead with "0". The following states of the cells are updated with a simple rule, i.e. a dead cell become alive if it is surrounded by three alive cells and an alive cell becomes dead if it is surrounded by three dead cells. It turns out that the game of life reveals a rich behaviour. This demonstrates the most important feature of the CA that extremely simple rules can build very complex system which cannot be extrapolated from the individual properties. In other words, with a suitable rules a CA can be used to simulate complex physical phenomena in the real world. In fact, many applications have shown that a CA is a very simple approach to complex physical phenomena [4, 5].

1.2.2 Lattice Gas Automata

A lattice gas automata is a particular class of the cellular automata. It is developed as a simple, fully discrete microscopic model for a fluid based on fictitious particles residing on a regular lattice. Such particles move one lattice unit in their directions of their velocities. Two or more particles arriving at the same site can collide. The important feature of the lattice gas automata is that mass and momentum are explicitly conserved which differ from a cellular automata. This is a very desirable feature in simulating real physical problems. In fact, it can be shown that the summations of the microdynamic mass and momentum equations can be asymptotically equivalent to the Navier-Stokes equation for incompressible flows.

In 1976 Hardy et al. [6] proposed the first fully discrete model for a fluid on a square lattice (HPP model). It was shown that the model simulate flow equations. Since the model is built on the 4-speed square lattice and particles move only in four directions, i.e. east, north, west and south, the resultant flow equations were anisotropic. Ten years later, Frisch et al. [7] found the fact that the symmetry of the lattice plays a dominant role in recovery of the Navier-Stokes equation. Based on a 6-speed hexagonal lattice they first obtained a correct lattice gas automata (FHP model) for Navier-Stokes equation.

The LGA comprises two steps: streaming and collision. The former is represented by a lattice pattern and the later by a collision operator. On macroscopic level in physics, these two steps simulate convection and diffusion phenomena, respectively. Consequently, a lattice pattern and collision operator determine the basic feature of a LGA. There are several models developed for these in the literature [8].

The equation for the LGA is

$$n_\alpha(\mathbf{x}+\mathbf{e}_\alpha, t+1) = n_\alpha(\mathbf{x}, t) + \Omega_\alpha[n(\mathbf{x}, t)], \qquad \alpha = 0, 1, ..., M, \tag{1.1}$$

where n_α is a Boolean variable that is used as an indication of the presence with $n_\alpha = 1$ or absence with $n_\alpha = 0$ of particles, $\mathbf{e}_\alpha$ is the local constant particle velocity, Ω_α is the collision operator, and M is the number of directions of the particle velocities.

The physical variables, density and velocities are defined by

$$\rho = \sum_{\alpha=0}^{M} <n_\alpha>, \qquad u_i = \frac{1}{\rho}\sum_{\alpha=0}^{M} <n_\alpha> e_{\alpha i}, \tag{1.2}$$

in which $<n_\alpha>$ denotes the ensemble average of n_α in statistical physics.

Often, simulations generated with a LGA are very noisy due to its Boolean nature [2]. Also, the numerical procedure involves probabilities which reduces the efficiency of a LGA. This leads to the birth of the lattice Boltzmann method.

1.2.3 Lattice Boltzmann Method

The lattice Boltzmann method is evolved from the LGA to overcome its difficulties. Its fundamental difference from the LGA is in that the Boolean variable is replaced by particle distribution functions, i.e. $<n_\alpha>= f_\alpha$ ($f_\alpha \geq 0$). If individual particle motion and particle - particle correlations are neglected, Eq. (1.1) can be replaced by the following lattice Boltzmann equation [9],

$$f_\alpha(\mathbf{x}+\mathbf{e}_\alpha, t+1) = f_\alpha(\mathbf{x}, t) + \Omega_\alpha[f(\mathbf{x}, t)], \qquad \alpha = 0, 1, ..., M. \tag{1.3}$$

Such approach eliminates the statistical noise in a LGA and retain all the advantages of locality in the kinetic form of a LGA [10].

Over the last few years, the study on the lattice Boltzmann method for fluid flows has received lots of attention, greatly improving and developing the method. McNamara and Zanetti [9] first used the lattice Boltzmann method as an alternative to the LGA in 1988. Since the collision operator takes a complex form, Higuera and Jiménez [11] made a first simplification for it. They linearized the collision term around its local equilibrium state. Later, several researchers [12, 13] suggested a simple linearized form for the collision operator by using a single time relaxation towards the local equilibrium, which is the so-called Bhatnagar-Gross-Krook [14] collision operator. This makes the

LBM become a very efficient and flexible method for simulating fluid flows. In fact, the lattice Boltzmann equation with the BGK collision operator is the most common version of the lattice Boltzmann method in present use.

1.3 Advantages

Apparently, the lattice Boltzmann method is a modern numerical techniques for solution of flow equations. Compared with the traditional computational fluid dynamics such as finite element method, finite difference method, finite volume method, it solves microscopic equations (lattice Boltzmann equation) and the density, velocity can be recovered from the macroscopic properties. Only simple arithmetic calculations can generate accurate solutions to the complex partial differential equations, flow equations. It is indeed an amazing method! What is more is that the lattice Boltzmann method provides an easy way to simulate complicated flows which is still a challenge to a traditional numerical method, e.g. multi-phase flows and flows through porous media. It has shown that the lattice Boltzmann method is a very promising computational method with a potential capability for simulating fluid flows in different areas [10]. The method is becoming a very powerful design tool in fluids engineering. The main advantages of the lattice Boltzmann method may be summarised as follows:

1. it consists of simple arithmetic calculations, hence it is easy to program;
2. only one single variable, the microscopic distribution function, is unknown and needs to be determined, which is particularly superior to a direct solution of flow equations with special treatments for the convection term and the pressure (depth) computations;
3. the current value of the distribution function depends only on the previous conditions which is ideal for parallel computations;
4. it is suitable for flows in complex geometry such as flows through porous media because of easy implementation of boundary conditions; and
5. it is easy to simulate complex flows, for example, multiphase flows and flows with variations of boundaries.

1.4 Objectives

Recently, the author developed the lattice Boltzmann models for the shallow water equations (LABSWE and LABSWE^{TM}) including turbulence modelling [15, 16]. It has shown that the LABSWE and LABSWE^{TM} can produce accurate solutions to the shallow water equations. Later, he proposed an elastic-collision scheme to achieve slip and semi-slip boundary conditions [17]. In addition, he introduced a centred scheme for accurate calculation of force terms in the lattice Boltzmann equations [18]. All of these immediately enable the

LABSWE and LABSWE^{TM} to be applied for solving the practical shallow water flow problems. Therefore, the present book is written timely in order to introduce this modern numerical method into research areas, industry sectors and educational regions.

The book aims to provide a simple, accurate and efficient numerical method for hydrodynamics and environmental hydraulics. The reader who has basic knowledge in algebra and calculus can understand. The method described itself will bring the reader to the present-day research frontier of the subject. The book can be used as a reference for scientists and engineers and also a text book for graduate students.

2

Shallow Water Flows

2.1 Introduction

Fluid flows obey conservation laws such as conservations of mass and momentum. Such conservation laws can be used to obtain a set of differential equations, namely, the continuity equation and the Navier-Stokes equation for description of fluid flows. These equations form a mathematical model for general fluid flows which are introduced in this chapter. Since a full solution of the Navier-Stokes equation for turbulent flows is generally beyond the power of the present computer, the subgrid-scale stress model is incorporated into the flow equations for turbulent flows. Based on the governing equations for general fluid flows, the shallow water equations are derived in detail, which is the mathematical model for shallow water flows.

2.2 General Flow Equations

The governing equations for incompressible flows are the 3D continuity and Navier-Stokes equations. They are derived according to the mass conservation and the Newton's second law of motion, which can be written in tensor form for the continuity equation,

$$\frac{\partial u_j}{\partial x_j} = 0 \tag{2.1}$$

and the momentum equation,

$$\frac{\partial u_i}{\partial t} + \frac{\partial (u_i u_j)}{\partial x_j} = f_i - \frac{1}{\rho}\frac{\partial p}{\partial x_i} + \nu \frac{\partial^2 u_i}{\partial x_j \partial x_j}, \tag{2.2}$$

where the subscripts i and j are space direction indices and the Einstein summation convention is used; f_i the body force per unit mass in i direction; ρ the fluid density; t the time; p the pressure; and ν the kinematic viscosity.

In the Einstein summation convention, the repeated indices mean a summation over the space coordinates. For example, $\partial u_j / \partial x_j$ is the short notation of the following expression,

$$\frac{\partial u_j}{\partial x_j} = \frac{\partial u}{\partial x} + \frac{\partial v}{\partial y} + \frac{\partial w}{\partial z}, \tag{2.3}$$

in which, x_j is the Cartesian coordinate shown in Fig. 2.1, taking x, y and z in turn; u_j is the velocity component which takes u, v and w corresponding to that in x, y and z directions, respectively.

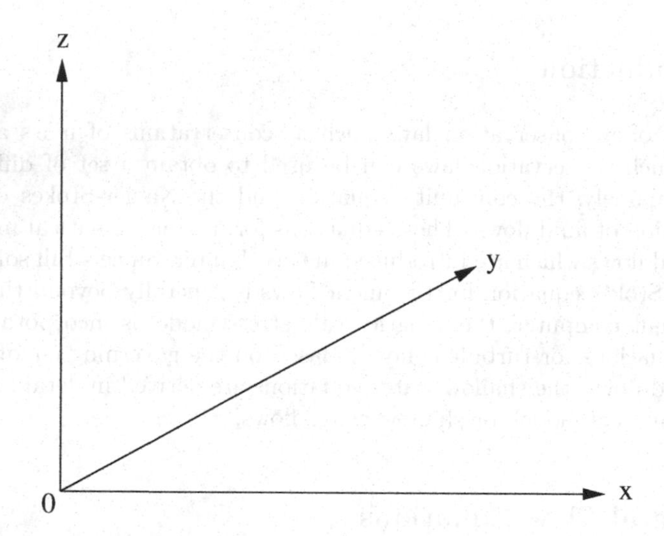

Fig. 2.1. Cartesian coordinate system: x-y stands for horizontal plane and z for vertical direction.

All the terms in the Navier-Stokes equation have physical interpretations. The whole left hand side of Eq. (2.2) is an inertia term in which $\partial(u_i u_j)/\partial x_j$ is also called convective term. The first term on the right hand side is the body force term, the second is the pressure term and the last is the viscous term. The equations (2.1) and (2.2) are the general governing equations for fluid flows. Except for a few simple situations, there is no analytical solution to them. With the development of the computer power, it is possible to obtain numerical solutions to the equations. This is the reason why a numerical method plays more and more important role in solving flow problems.

Theoretically, the Navier-Stokes equation can be used to solve turbulent flows in detail by means of a direct numerical simulation (DNS). A recent study has indicated that it is currently possible to carry out direct numerical simulations of homogeneous flows only at low Reynolds numbers because the requirement of a huge number of grid points in the solution domain is beyond modern computer technology [19]. The use of the Navier-Stokes equation for direct solutions to turbulent flow problems is still a long way in the future. In practice, an alternative way is often used to model flow turbulence by means of the modified Navier-Stokes equation in numerical methods. This can provide resolved scale property (space-filtered quantities) or mean property (time-averaged quantities). But none of them can predict the whole details of flow turbulence. Therefore, how to simulate turbulent flows efficiently and accurately is an active research area.

2.3 Subgrid-Scale Stress Model

Although the flows encountered are mostly turbulent in a natural river or a channel, only a few mean or resolved scale properties of turbulent flows are of primary importance from the point of view of engineering; hence the modified Navier-Stokes equation are widely applied to solve turbulent flow problems. This may be classified as two main categories: (1) time-averaged Navier-Stokes equation, the Reynolds equation, together with a model for the turbulence stress such as k-ϵ model; and (2) space-filtered Navier-Stokes equation, the large eddy simulation (LES), with subgrid-scale stress model for the unresolved scale stress. Recent research indicates that, although it is more expensive to use space-filtered Navier-Stokes equation than the Reynolds equation, the former will produce more accurate solution to turbulent flows and also provide considerable detailed characteristics of turbulent flows such as fluctuation of a physical quantity [20]. By contrast the latter can provide time-averaged features of flow turbulence but loses all instantaneous characteristics which are basic turbulence features such as velocity fluctuations. Therefore, the large eddy simulation is used for turbulent flows in this book.

The flow equations for the large eddy simulation can be derived by introducing a space-filtered quantity in the continuity (2.1) and the momentum equation (2.2). The resulting space-filtered continuity and Navier-Stokes equation can be written as

$$\frac{\partial \tilde{u}_j}{\partial x_j} = 0 \tag{2.4}$$

and

$$\frac{\partial \tilde{u}_i}{\partial t} + \frac{\partial (\tilde{u}_i \tilde{u}_j)}{\partial x_j} = f_i - \frac{1}{\rho}\frac{\partial p}{\partial x_i} + \nu \frac{\partial^2 \tilde{u}_i}{\partial x_j \partial x_j} - \frac{\partial \tau_{ij}}{\partial x_j}, \tag{2.5}$$

where $\tilde{u}$ is the space-filtered velocity component in i direction defined by

$$\tilde{u}(x,y,z,t) = \int\int\int_{\Delta x \Delta y \Delta z} u(x,y,z,t) G(x,y,z,x',y',z') dx'dy'dz' \tag{2.6}$$

with a spatial filter function G. τ_{ij} is called the subgrid-scale stress (SGS) which reflects the effects of the unresolved scales with the resolved scales, i.e.

$$\tau_{ij} = \widetilde{u_i u_j} - \tilde{u}_i \tilde{u}_j. \tag{2.7}$$

By following the Boussinesq assumption for turbulent stresses, we may further represent the subgrid-scale stress with an SGS eddy viscosity ν_e as

$$\tau_{ij} = -\nu_e \left(\frac{\partial \tilde{u}_i}{\partial x_j} + \frac{\partial \tilde{u}_j}{\partial x_i} \right). \tag{2.8}$$

Substitution of Eq. (2.8) into Eq. (2.5) leads to the following momentum equation,

$$\frac{\partial \tilde{u}_i}{\partial t} + \frac{\partial (\tilde{u}_i \tilde{u}_j)}{\partial x_j} = f_i - \frac{1}{\rho} \frac{\partial p}{\partial x_i} + (\nu + \nu_e) \frac{\partial^2 \tilde{u}_i}{\partial x_j \partial x_j}. \tag{2.9}$$

In the book, the standard Smagorinsky SGS model [21] is used and the eddy viscosity ν_e is defined by

$$\nu_e = (C_s l_s)^2 \sqrt{S_{ij} S_{ij}}, \tag{2.10}$$

in which C_s is Smagorinsky constant, l_s is the characteristic length scale and S_{ij} is the magnitude of the large scale strain-rate tensor,

$$S_{ij} = \frac{1}{2} \left(\frac{\partial \tilde{u}_i}{\partial x_j} + \frac{\partial \tilde{u}_j}{\partial x_i} \right). \tag{2.11}$$

The equations (2.4) and (2.9) are the modified continuity and Navier-Stokes equation used as large eddy simulations for turbulent flows. The resolved large scale fields are numerically solved and the effect of unresolved scale eddies is modelled with a subgrid-scale stress model. The finer the grid size, the less the unresolved scale eddies. If the size is small enough to resolve all the eddy scales, the numerical simulation becomes a direct numerical simulation, which is of course impossible for most turbulent flows at present or in the near future as indicated in Section 2.2. Since the lattice spacing, mesh size, in the lattice Boltzmann method is usually much smaller than that used in a traditional computational method for fluid flows, it is expected that an incorporation of a subgrid-scale stress model into the lattice Boltzmann method can produce more accurate solutions to turbulent flows.

2.4 Shallow Water Equations

The water depth in flows existing in rivers, channels, coastal areas, estuaries and harbours is usually much smaller than the horizontal scale. Such flows are characterised by horizontal motions. In a mathematical model, the assumption of the hydrostatic pressure is often used to replace the momentum equation in the vertical direction; hence the vertical acceleration is ignored. The flows can be described with either 3D shallow water equations or 2D shallow water equations. In the former situations, the vertical velocity is calculated from the continuity equations, while in the latter a depth-averaged quantity is used without the vertical velocity. None of a model based on either 3D shallow water equations or 2D shallow water equations can predict vertical separation or predict it accurately [22]. This suggests that the 3D shallow water model has the same weakness as the 2D shallow water one. As a result, the 2D shallow water equations are widely used as a mathematical model for shallow water flows, which are used in this book.

The general 2D governing equations for shallow water flows can be derived based on the general flow equations (2.1) and (2.2). For water flows on the earth, there are two body forces: gravity in vertical direction and Coriolis acceleration in horizontal plane due to the earth's rotation [23]. Using the coordinate system shown in Fig. 2.1, we have the following body forces,

$$f_x = f_c v, \qquad f_y = -f_c u, \qquad f_z = -g, \tag{2.12}$$

where $g = 9.81\ m/s^2$ is the gravitational acceleration; u and v are the velocity components in x and y directions, respectively; $f_c = 2\omega \sin\phi$ is the Coriolis parameter in which $\omega \approx 7.3 \times 10^{-5}\ rad/s$ is the earth's rotation and ϕ is the earth's latitude at the site of interest.

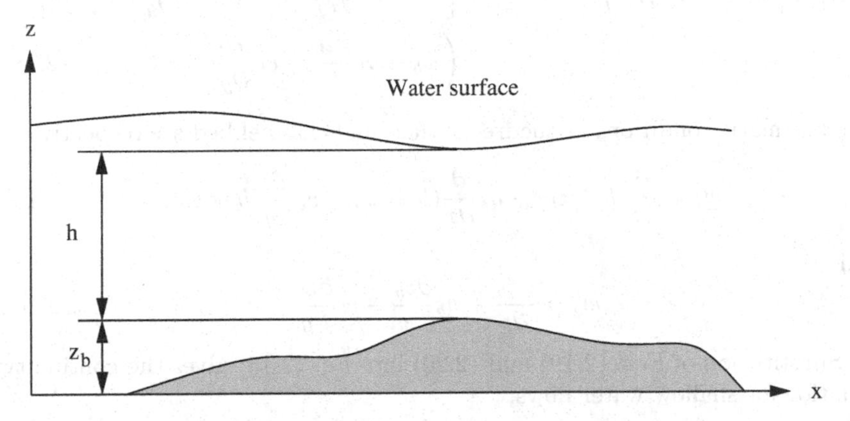

Fig. 2.2. Sketch for Derivation of Shallow Water Equations

In order to determine the continuity equation in terms of depth-averaged quantities, we integrate Eq. (2.1) over depth and obtain

$$\int_{z_b}^{h+z_b} \left(\frac{\partial u}{\partial x} + \frac{\partial v}{\partial y} + \frac{\partial w}{\partial z} \right) dz = 0, \tag{2.13}$$

which leads to

$$\int_{z_b}^{h+z_b} \frac{\partial u}{\partial x} dz + \int_{z_b}^{h+z_b} \frac{\partial v}{\partial y} dz + w_s - w_b = 0, \tag{2.14}$$

where w_s and w_b is the vertical velocities at the free surface and channel bed, respectively; h is the water depth; and z_b is the bed elevation above datum (see Fig. 2.2).

By using the Leibnitz rule [24],

$$\int_a^b \frac{\partial f(x,y)}{\partial y} dx = \frac{\partial}{\partial y} \int_a^b f(x,y)dx - f(b,y)\frac{\partial b}{\partial y} + f(a,y)\frac{\partial a}{\partial y}, \tag{2.15}$$

we can write the first term on the left hand side of Eq. (2.14) as

$$\int_{z_b}^{h+z_b} \frac{\partial u}{\partial x} dz = \frac{\partial}{\partial x} \int_{z_b}^{h+z_b} u dz - u_s \frac{\partial}{\partial x}(h+z_b) + u_b \frac{\partial z_b}{\partial x}, \tag{2.16}$$

and the second as

$$\int_{z_b}^{h+z_b} \frac{\partial v}{\partial y} dz = \frac{\partial}{\partial y} \int_{z_b}^{h+z_b} v dz - v_s \frac{\partial}{\partial y}(h+z_b) + v_b \frac{\partial z_b}{\partial y}. \tag{2.17}$$

Substitution of Eqs. (2.16) and (2.17) into (2.14) results in

$$\frac{\partial}{\partial x} \int_{z_b}^{h+z_b} u dz + \frac{\partial}{\partial y} \int_{z_b}^{h+z_b} v dz + \left[w_s - u_s \frac{\partial}{\partial x}(h+z_b) - v_s \frac{\partial}{\partial y}(h+z_b) \right] - \left(w_b - u_b \frac{\partial z_b}{\partial x} - v_b \frac{\partial z_b}{\partial y} \right) = 0. \tag{2.18}$$

The kinematic conditions at the free surface and channel bed are respectively,

$$w_s = \frac{\partial}{\partial t}(h+z_b) + u_s \frac{\partial}{\partial x}(h+z_b) + v_s \frac{\partial}{\partial y}(h+z_b), \tag{2.19}$$

and

$$w_b = \frac{\partial z_b}{\partial t} + u_b \frac{\partial z_b}{\partial x} + v_b \frac{\partial z_b}{\partial y}. \tag{2.20}$$

Substitution of Eqs. (2.19) and (2.20) into Eq. (2.18) gives the continuity equation for shallow water flows,

$$\frac{\partial h}{\partial t} + \frac{\partial (h\overline{u})}{\partial x} + \frac{\partial (h\overline{v})}{\partial y} = 0, \tag{2.21}$$

where $\overline{u}$ and $\overline{v}$ are depth-averaged velocities and defined as

$$\overline{u} = \frac{1}{h} \int_{z_b}^{h+z_b} u dz, \qquad \overline{v} = \frac{1}{h} \int_{z_b}^{h+z_b} v dz. \tag{2.22}$$

For the momentum equation, we integrate the x component of (2.2) over depth and obtain

$$\int_{z_b}^{h+z_b} \left[\frac{\partial u}{\partial t} + \frac{\partial (uu)}{\partial x} + \frac{\partial (vu)}{\partial y} + \frac{\partial (wu)}{\partial z} \right] dz = \int_{z_b}^{h+z_b} f_c v dz + \int_{z_b}^{h+z_b} \left[-\frac{1}{\rho}\frac{\partial p}{\partial x} + \nu \left(\frac{\partial^2 u}{\partial x^2} + \frac{\partial^2 u}{\partial y^2} + \frac{\partial^2 u}{\partial z^2} \right) \right] dz. \tag{2.23}$$

Again, using the Leibnitz rule (2.15) for the first three terms on the left hand side of the above equation results in

$$\int_{z_b}^{h+z_b} \frac{\partial u}{\partial t} dz = \frac{\partial}{\partial t} \int_{z_b}^{h+z_b} u dz - u_s \frac{\partial}{\partial t}(h+z_b) + u_b \frac{\partial z_b}{\partial t}, \tag{2.24}$$

$$\int_{z_b}^{h+z_b} \frac{\partial (uu)}{\partial x} dz = \frac{\partial}{\partial x} \int_{z_b}^{h+z_b} uu dz - u_s u_s \frac{\partial}{\partial x}(h+z_b) + u_b u_b \frac{\partial z_b}{\partial x}, \tag{2.25}$$

and

$$\int_{z_b}^{h+z_b} \frac{\partial (vu)}{\partial y} dz = \frac{\partial}{\partial y} \int_{z_b}^{h+z_b} vu dz - v_s u_s \frac{\partial}{\partial y}(h+z_b) + v_b u_b \frac{\partial z_b}{\partial y}. \tag{2.26}$$

Obviously, the last term on the left hand side of Eq. (2.23) can be directly integrated as

$$\int_{z_b}^{h+z_b} \frac{\partial (wu)}{\partial z} dz = w_s u_s - w_b u_b. \tag{2.27}$$

After putting Eqs. (2.24) - (2.27) together and rearrange them, we obtain

$$\begin{aligned} \int_{z_b}^{h+z_b} \left[\frac{\partial u}{\partial t} + \frac{\partial (uu)}{\partial x} + \frac{\partial (vu)}{\partial y} + \frac{\partial (wu)}{\partial z} \right] dz = \\ \frac{\partial}{\partial t} \int_{z_b}^{h+z_b} u dz + \frac{\partial}{\partial x} \int_{z_b}^{h+z_b} uu dz + \frac{\partial}{\partial y} \int_{z_b}^{h+z_b} vu dz \\ + u_s \left[w_s - \frac{\partial}{\partial t}(h+z_b) - u_s \frac{\partial}{\partial x}(h+z_b) - v_s \frac{\partial}{\partial y}(h+z_b) \right] \\ - u_b \left(w_b - \frac{\partial z_b}{\partial t} - u_b \frac{\partial z_b}{\partial x} - v_b \frac{\partial z_b}{\partial y} \right). \end{aligned} \tag{2.28}$$

With reference to the kinematic conditions (2.19) and (2.20) together with the definition (2.22), the above expression can be simplified as

$$\begin{aligned} \int_{z_b}^{h+z_b} \left[\frac{\partial u}{\partial t} + \frac{\partial (uu)}{\partial x} + \frac{\partial (vu)}{\partial y} + \frac{\partial (wu)}{\partial z} \right] dz = \\ \frac{\partial (\bar{u} h)}{\partial t} + \frac{\partial}{\partial x} \int_{z_b}^{h+z_b} uu dz + \frac{\partial}{\partial y} \int_{z_b}^{h+z_b} vu dz. \end{aligned} \tag{2.29}$$

By using the second mean value theorem for integrals [25],

$$\int_a^b f(x) g(x) dx = f(\zeta) \int_a^b g(x) dx, \tag{2.30}$$

we can express the second term of the left hand side of Eq. (2.29) as

$$\int_{z_b}^{h+z_b} uudz = \breve{u}_1 \int_{z_b}^{h+z_b} udz = \breve{u}_1 h\overline{u}, \tag{2.31}$$

and the last term as

$$\int_{z_b}^{h+z_b} vudz = \breve{u}_2 \int_{z_b}^{h+z_b} vdz = \breve{u}_2 h\overline{v}. \tag{2.32}$$

Apparently, the use of the second mean value theorem (2.30) implies that the horizontal velocities at time t, $u(x,y,z,t)$ and $v(x,y,z,t)$, do not change their directions over the water depth, i.e. $u(x,y,z,t)$ is always either $u(x,y,z,t) \geq 0$ from channel bed to free surface at the horizontal location (x,y) or $u(x,y,z,t) < 0$ from channel bed to free surface, and so is $v(x,y,z,t)$. This mathematically leads to the reason why a model based on 2D shallow water equations pre-excludes flow separations in vertical direction.

If $\breve{u}_1 = \theta_1 \overline{u}$ and $\breve{u}_2 = \theta_2 \overline{u}$ are assumed, inserting Eqs. (2.31) and (2.32) into Eq. (2.29) results in

$$\int_{z_b}^{h+z_b} \left[\frac{\partial u}{\partial t} + \frac{\partial (uu)}{\partial x} + \frac{\partial (vu)}{\partial y} + \frac{\partial (wu)}{\partial z} \right] dz = \frac{\partial (h\overline{u})}{\partial t} + \frac{\partial (\theta_1 h\overline{u}\,\overline{u})}{\partial x} + \frac{\partial (\theta_2 h\overline{v}\,\overline{u})}{\partial y}, \tag{2.33}$$

where θ_1 and θ_2 are called momentum correction factors which can be determined based on Eqs. (2.31) and (2.32) as

$$\theta_1 = \frac{1}{h\overline{u}\,\overline{u}} \int_{z_b}^{h+z_b} uudz, \qquad \theta_2 = \frac{1}{h\overline{v}\,\overline{u}} \int_{z_b}^{h+z_b} vudz. \tag{2.34}$$

Similarly, we can obtain the following expression for the terms on the left hand side of the momentum equation (2.2) in y direction,

$$\int_{z_b}^{h+z_b} \left[\frac{\partial v}{\partial t} + \frac{\partial (uv)}{\partial x} + \frac{\partial (vv)}{\partial y} + \frac{\partial (wv)}{\partial z} \right] dz = \frac{\partial (\overline{v}h)}{\partial t} + \frac{\partial (\theta_2 h\overline{u}\,\overline{v})}{\partial x} + \frac{\partial (\theta_3 h\overline{v}\,\overline{v})}{\partial y}, \tag{2.35}$$

with an additional momentum correction factor θ_3 defined by

$$\theta_3 = \frac{1}{h\overline{v}\,\overline{v}} \int_{z_b}^{h+z_b} vvdz. \tag{2.36}$$

The first term, Coriolis force, on the right hand side of Eq. (2.23) can be integrated as

$$\int_{z_b}^{h+z_b} f_c vdz = f_c h\overline{v}. \tag{2.37}$$

The second term related to the pressure p can be derived from the momentum equation. Since in shallow water flows the vertical acceleration becomes unimportant compared with the horizontal effect, the momentum equation (2.2) in the vertical direction is reduced with $w \approx 0$ to

$$\frac{\partial p}{\partial z} = -\rho g, \tag{2.38}$$

which can be integrated, giving

$$p = -\rho g z + C_0, \tag{2.39}$$

in which C_0 is called the integration constant.

By using the boundary condition that the pressure at the free surface is the atmospheric pressure p_a, i.e. $p = p_a$ when $z = h + z_b$, in the above equation, we have

$$C_0 = \rho g(h + z_b) + p_a. \tag{2.40}$$

Substitution of the above expression into Eq. (2.39) leads to

$$p = \rho g(h + z_b - z) + p_a. \tag{2.41}$$

Since the difference in atmospheric pressure at water surface is often insignificant in most coastal, estuarine, hydraulic engineering [23], p_a is almost constant in the modelling area and it is normally set to zero, i.e. $p_a = 0$. Hence Eq. (2.41) becomes

$$p = \rho g(h + z_b - z). \tag{2.42}$$

This is usually referred to the hydrostatic pressure approximation in shallow water flows. Differentiating Eq. (2.42) with respect to x results in

$$\frac{\partial p}{\partial x} = \rho g \frac{\partial}{\partial x}(h + z_b). \tag{2.43}$$

Obviously, neither the water depth h nor the bed height z_b is dependent on vertical direction. They are functions of the horizontal coordinates x and y only, indicating $\partial p/\partial x$ is a function of x and y only; hence we have

$$\int_{z_b}^{h+z_b} \frac{1}{\rho}\frac{\partial p}{\partial x} dz = \frac{h}{\rho}\frac{\partial p}{\partial x}. \tag{2.44}$$

Substitution of Eq. (2.43) into above equation leads to

$$\int_{z_b}^{h+z_b} \frac{1}{\rho}\frac{\partial p}{\partial x} dz = gh\frac{\partial}{\partial x}(h + z_b). \tag{2.45}$$

Now, we introduce the following approximations,

$$\int_{z_b}^{h+z_b} \nu\frac{\partial^2 u}{\partial x^2} dz \approx \nu\frac{\partial^2 (h\overline{u})}{\partial x \partial x} \tag{2.46}$$

and

$$\int_{z_b}^{h+z_b} \nu \frac{\partial^2 u}{\partial y^2} dz \approx \nu \frac{\partial^2 (h\overline{u})}{\partial y \partial y}. \tag{2.47}$$

for the third and the fourth terms.

The last term on the right hand side of Eq. (2.23) can be calculated as

$$\int_{z_b}^{h+z_b} \nu \frac{\partial^2 u}{\partial z^2} dz = \left(\nu \frac{\partial u}{\partial z}\right)_s - \left(\nu \frac{\partial u}{\partial z}\right)_b. \tag{2.48}$$

Usually, the terms on the right hand side of the above expression can be approximated with the surface wind shear stress and the bed shear stress in x direction, respectively, i.e.

$$\left(\nu \frac{\partial u}{\partial z}\right)_s = \frac{\tau_{wx}}{\rho}, \qquad \left(\nu \frac{\partial u}{\partial z}\right)_b = \frac{\tau_{bx}}{\rho}. \tag{2.49}$$

Eq. (2.48) then becomes

$$\int_{z_b}^{h+z_b} \nu \frac{\partial^2 u}{\partial z^2} dz = \frac{\tau_{wx}}{\rho} - \frac{\tau_{bx}}{\rho}. \tag{2.50}$$

Substitution of Eqs. (2.33), (2.37), (2.45) - (2.47) and (2.50) into Eq. (2.23) results in the x momentum equation for shallow water flows,

$$\frac{\partial (h\overline{u})}{\partial t} + \frac{\partial (\theta_1 h\overline{u}\,\overline{u})}{\partial x} + \frac{\partial (\theta_2 h\overline{v}\,\overline{u})}{\partial y} = -g\frac{\partial}{\partial x}(\frac{h^2}{2}) + \nu \frac{\partial^2 (h\overline{u})}{\partial x \partial x} + \nu \frac{\partial^2 (h\overline{u})}{\partial y \partial y}$$
$$- gh\frac{\partial z_b}{\partial x} + f_c h\overline{v} + \frac{\tau_{wx}}{\rho} - \frac{\tau_{bx}}{\rho}. \tag{2.51}$$

Similarly, we can derive the momentum equation in the y direction for shallow water flows as

$$\frac{\partial (h\overline{v})}{\partial t} + \frac{\partial (\theta_2 h\overline{u}\,\overline{v})}{\partial x} + \frac{\partial (\theta_3 h\overline{v}\,\overline{v})}{\partial y} = -g\frac{\partial}{\partial y}(\frac{h^2}{2}) + \nu \frac{\partial^2 (h\overline{v})}{\partial x \partial x} + \nu \frac{\partial^2 (h\overline{v})}{\partial y \partial y}$$
$$- gh\frac{\partial z_b}{\partial y} - f_c h\overline{u} + \frac{\tau_{wy}}{\rho} - \frac{\tau_{by}}{\rho}. \tag{2.52}$$

Theoretically, the momentum correction factors θ_1, θ_2 and θ_3 can be calculated by Eqs. (2.34) and (2.36) if velocity profiles for u and v are assumed or known. However, there are no universal velocity profiles valid for flows in most situations, e.g. flows involving circulation/separation or in a channel with complex geometry, and then it is difficult to estimate these momentum correction factors θ_1, θ_2 and θ_3. Instead, $\theta_1 = 1$, $\theta_2 = 1$ and $\theta_3 = 1$ are widely used in numerical analogues for shallow water flows and it is found that they can provide a good approximation in most situations [26, 27, 28, 29].

Therefore, $\theta_1 = 1$, $\theta_2 = 1$ and $\theta_3 = 1$ are adopted, resulting in Eqs. (2.51) and (2.52) in the following forms,

$$\frac{\partial(h\overline{u})}{\partial t}+\frac{\partial(h\overline{u}\,\overline{u})}{\partial x}+\frac{\partial(h\overline{v}\,\overline{u})}{\partial y}=-g\frac{\partial}{\partial x}\left(\frac{h^2}{2}\right)+\nu\frac{\partial^2(h\overline{u})}{\partial x\partial x}+\nu\frac{\partial^2(h\overline{u})}{\partial y\partial y}$$
$$-gh\frac{\partial z_b}{\partial x}+f_c h\overline{v}+\frac{\tau_{wx}}{\rho}-\frac{\tau_{bx}}{\rho}, \tag{2.53}$$

and

$$\frac{\partial(h\overline{v})}{\partial t}+\frac{\partial(h\overline{u}\,\overline{v})}{\partial x}+\frac{\partial(h\overline{v}\,\overline{v})}{\partial y}=-g\frac{\partial}{\partial y}\left(\frac{h^2}{2}\right)+\nu\frac{\partial^2(h\overline{v})}{\partial x\partial x}+\nu\frac{\partial^2(h\overline{v})}{\partial y\partial y}$$
$$-gh\frac{\partial z_b}{\partial y}-f_c h\overline{u}+\frac{\tau_{wy}}{\rho}-\frac{\tau_{by}}{\rho}. \tag{2.54}$$

After all the overbars in the above two equations are dropped for convenience, the continuity equation (2.21) and the momentum equations (2.53) and (2.54) can be simply written in a tensor form as

$$\frac{\partial h}{\partial t}+\frac{\partial(hu_j)}{\partial x_j}=0, \tag{2.55}$$

$$\frac{\partial(hu_i)}{\partial t}+\frac{\partial(hu_iu_j)}{\partial x_j}=-g\frac{\partial}{\partial x_i}(\frac{h^2}{2})+\nu\frac{\partial^2(hu_i)}{\partial x_j\partial x_j}+F_i, \tag{2.56}$$

where F_i is called a force term in this book and defined as

$$F_i=-gh\frac{\partial z_b}{\partial x_i}+\frac{\tau_{wi}}{\rho}-\frac{\tau_{bi}}{\rho}+E_i, \tag{2.57}$$

with the Coriolis term E_i given by

$$E_i=\begin{cases} f_c hv, & i=x,\\ -f_c hu, & i=y.\end{cases} \tag{2.58}$$

The bed shear stress τ_{bi} in i direction is given by the depth-averaged velocities,

$$\tau_{bi}=\rho C_b u_i\sqrt{u_ju_j}, \tag{2.59}$$

in which C_b is the bed friction coefficient, which may be either constant or estimated from $C_b=g/C_z^2$, where C_z is the Chezy coefficient given with either Manning equation,

$$C_z=\frac{h^{\frac{1}{6}}}{n_b}, \tag{2.60}$$

here n_b is the Manning's coefficient at the bed, or the Colebrook-White equation [30],

$$C_z=-\sqrt{32g}\,\log_{10}\left(\frac{K_s}{14.8h}+\frac{1.255\nu C_z}{4\sqrt{2g}\;hu}\right), \tag{2.61}$$

with the Nikuradse equivalent sand roughness K_s.

The wind shear stress τ_{wi} is usually expressed as

$$\tau_{wi}=\rho_a C_w u_{wi}\sqrt{u_{wj}u_{wj}}, \tag{2.62}$$

where ρ_a is the density of air, C_w the resistance coefficient, and u_{wi} the component of the wind velocity in i direction.

2.5 Various Numerical Methods

In the literature, many computational methods are available for solutions of the shallow water equations (2.55) and (2.56). For example, a SIMPLE-like method [27, 31], a semi-implicit method [32], and a Godunov-type method [33, 34]. All of these methods are formulated based on direct solutions of the shallow water equations: i.e. the equations are firstly discretized by means of a finite approach such as finite difference method, finite volume method and finite element method to obtain a set of algebraic equations which are then solved with a numerical technique. The common weakness of the methods is that a treatment is required in the numerical procedures for either convective term, or depth computations, or numerical flux, or source terms. This undoubtedly introduces more or less uncertainty into the models.

In contrast, a completely different numerical method is developed for fluid flows over the last decade. It is called the lattice Boltzmann method, which is evolved from the lattice gas automata without the weaknesses associated with the lattice gas automata [10, 35]. As indicated in Section 1.3, this is a very promising computational method in simulating fluid flows [10]. Recently, the author [15, 16] has developed the lattice Boltzmann model for shallow water equations (LABSWE and LABSWE^{TM}), demonstrating that this is a simple, efficient and accurate model for shallow water flows with or without turbulence effect. The LABSWE and LABSWE^{TM} form the core of this book.

2.6 Closure

The Navier-Stokes equation is a general mathematical model for fluid flows. However, its direct solution to most turbulent flows is impossible based on present computer power. In practice, either a space-filtered or a time-averaged Navier-Stokes equations is used for simulation of turbulent flows. The space-filtered Navier-Stokes equations is used in this book due to its accuracy. As long as a vertical acceleration is not important, the general flow equations can be simplified to the shallow water equations which provide a simpler and more efficient mathematical model for shallow water flows. Although various numerical methods for the shallow water equations are available, the book aims to describe the lattice Boltzmann method for the shallow water equations because of its potential capability.

3

Lattice Boltzmann Method

3.1 Introduction

The lattice Boltzmann method is a discrete computational method based upon the lattice gas automata - a simplified, fictitious molecular model. It consists of three basic tasks: lattice Boltzmann equation, lattice pattern and local equilibrium distribution function. The former two are standard, which is the same for fluid flows. The latter determines what flow equations are solved by the lattice Boltzmann model, which is often derived for certain flow equations such as the equations for shallow water flows. These tasks are described in this chapter and some related aspects are also discussed.

3.2 Lattice Boltzmann Equation

The lattice Boltzmann method (LBM) is originally evolved from the LGA, i.e. the equation for the LGA is replaced with the lattice Boltzmann equation (LBE). As will be shown later in Section 3.8, the LBE can effectively be derived from the continuum Boltzmann equation [10], leading it to be self-explanatory in statistical physics. It is generally valid for fluid flows such as shallow water flows. According to the origin of the lattice Boltzmann method, it consists of two steps: a streaming step and a collision step. In the streaming step, the particles move to the neighbouring lattice points in their directions of their velocities, which is governed by

$$f_\alpha(\mathbf{x} + \mathbf{e}_\alpha \Delta t, t + \Delta t) = f'_\alpha(\mathbf{x}, t) + \frac{\Delta t}{N_\alpha e^2} e_{\alpha i} F_i(\mathbf{x}, t), \tag{3.1}$$

where f_α is the distribution function of particles; f'_α is the value of f_α before the streaming; $e = \Delta x/\Delta t$; Δx is the lattice size; Δt is the time step; F_i is the component of the force in i direction; $\mathbf{e}_\alpha$ is the velocity vector of a particle in the α link and N_α is a constant, which is decided by the lattice pattern as

$$N_\alpha = \frac{1}{e^2}\sum_\alpha e_{\alpha i} e_{\alpha i}. \tag{3.2}$$

In the collision step, the arriving particles at the points interact one another and change their velocity directions according to scattering rules, which is expressed as

$$f'_\alpha(\mathbf{x},t) = f_\alpha(\mathbf{x},t) + \Omega_\alpha[f(\mathbf{x},t)], \tag{3.3}$$

in which Ω_α is the collision operator which controls the speed of change in f_α during collision.

Theoretically, Ω_α is generally a matrix, which is decided by the microscopic dynamics. Higuera and Jiménez [11] first introduced an idea to linearize the collision operator around its local equilibrium state. This greatly simplifies the collision operator. Based on this idea, Ω_α can be expanded about its equilibrium value [36],

$$\Omega_\alpha(f) = \Omega_\alpha(f^{eq}) + \frac{\partial \Omega_\alpha(f^{eq})}{\partial f_\beta}(f_\beta - f_\beta^{eq}) + O[(f_\beta - f_\beta^{eq})^2], \tag{3.4}$$

where f_α^{eq} is the local equilibrium distribution function.

The solution process of the lattice Boltzmann equation is characterized by $f_\beta \rightarrow f_\beta^{eq}$, implying $\Omega_\alpha(f^{eq}) \approx 0$. After the higher-order terms in Eq. (3.4) are neglected, we obtain a linearized collision operator,

$$\Omega_\alpha(f) \approx \frac{\partial \Omega_\alpha(f^{eq})}{\partial f_\beta}(f_\beta - f_\beta^{eq}). \tag{3.5}$$

If assuming the local particle distribution relaxes to an equilibrium state at a single rate τ [12, 13],

$$\frac{\partial \Omega_\alpha(f^{eq})}{\partial f_\beta} = -\frac{1}{\tau}\delta_{\alpha\beta}, \tag{3.6}$$

where $\delta_{\alpha\beta}$ is the Kronecker delta function,

$$\delta_{\alpha\beta} = \begin{cases} 0, & \alpha \neq \beta, \\ 1, & \alpha = \beta, \end{cases} \tag{3.7}$$

we can write Eq. (3.5) as

$$\Omega_\alpha(f) = -\frac{1}{\tau}\delta_{\alpha\beta}(f_\beta - f_\beta^{eq}), \tag{3.8}$$

resulting in the lattice BGK collision operator [14],

$$\Omega_\alpha(f) = -\frac{1}{\tau}(f_\alpha - f_\alpha^{eq}), \tag{3.9}$$

and τ is called as the single relaxation time. This makes the lattice Boltzmann equation extremely simple and efficient; hence it is widely used in a lattice

Boltzmann model for fluid flows. With reference to Eq. (3.9), the streaming and collision steps are usually combined into the following lattice Boltzmann equation,

$$f_\alpha(\mathbf{x} + \mathbf{e}_\alpha \Delta t, t + \Delta t) - f_\alpha(\mathbf{x}, t) = -\frac{1}{\tau}(f_\alpha - f_\alpha^{eq}) + \frac{\Delta t}{N_\alpha e^2} e_{\alpha i} F_i, \tag{3.10}$$

which is the most popular form of the lattice Boltzmann equation in use today.

3.3 Lattice Pattern

Lattice pattern in the lattice Boltzmann method has two functions: representing grid points and determining particles' motions. The former plays a similar role to that in the traditional numerical methods. The latter defines a microscopic model for molecular dynamics. In addition, the constant N_α in Eq. (3.10) is determined by the lattice pattern.

In 2D situations, there are generally two types of lattice patterns: square lattice and hexagonal lattice suggested in the literature. Their examples are shown in Figs. 3.1 and 3.2, respectively. Depending to the number of particle speed at lattice node, the square lattice can have 4-speed, 5-speed, 8-speed or 9-speed models, and the hexagonal lattice can have 6-speed and 7-speed models. Not all of these models have sufficient lattice symmetry which is a dominant requirement for recovery of the correct flow equations [7]. Theoretical analysis and numerical studies indicate that both 9-speed square lattice (see Fig. 3.1) and 7-speed hexagonal lattice (see Fig. 3.2) have such property and satisfactory performance in numerical simulations. They are widely used in the lattice Boltzmann methods. However, the recent study has shown that the 9-speed square lattice usually gives more accurate results than that based on hexagonal lattice [37]. Furthermore, the use of the square lattice provides an easy way to implement different boundary conditions [17], e.g. only by using the square lattice, can the force term associated with a gradient and boundary conditions be accurately and easily determined. Hence the 9-speed square lattice is preferred in use today and it is applied in this book throughout (see Appendix A for a lattice Boltzmann model on the 7-speed hexagonal lattice for shallow water flows).

On the 9-speed square lattice shown in Fig. 3.1, each particle moves one lattice unit at its velocity only along the eight links indicated with 1 - 8, in which 0 indicates the rest particle with zero speed. The velocity vector of particles is defined by

$$\mathbf{e}_\alpha = \begin{cases} (0,0), & \alpha = 0, \\ e\left[\cos\frac{(\alpha-1)\pi}{4}, \sin\frac{(\alpha-1)\pi}{4}\right], & \alpha = 1,3,5,7, \\ \sqrt{2}e\left[\cos\frac{(\alpha-1)\pi}{4}, \sin\frac{(\alpha-1)\pi}{4}\right], & \alpha = 2,4,6,8. \end{cases} \tag{3.11}$$

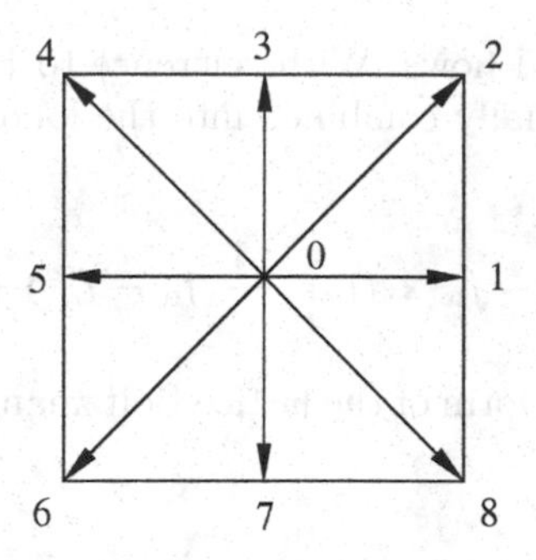

Fig. 3.1. 9-speed square lattice.

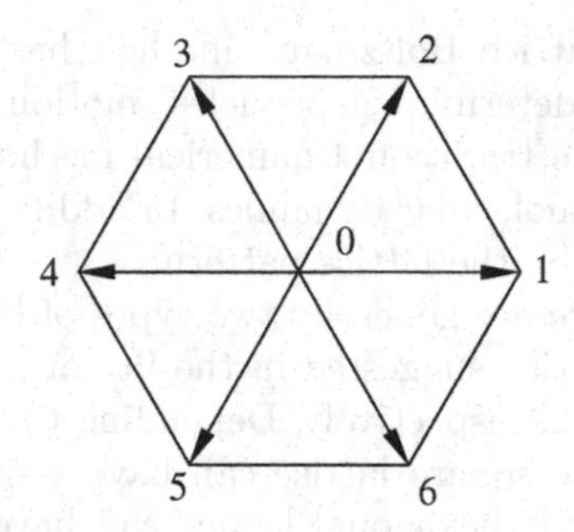

Fig. 3.2. 7-speed hexagonal lattice.

It is easy to show that the 9-speed square lattice has the following basic features,

$$\sum_\alpha e_{\alpha i} = \sum_\alpha e_{\alpha i} e_{\alpha j} e_{\alpha k} = 0, \tag{3.12}$$

$$\sum_\alpha e_{\alpha i} e_{\alpha j} = 6e^2 \delta_{ij}, \tag{3.13}$$

$$\sum_\alpha e_{\alpha i} e_{\alpha j} e_{\alpha k} e_{\alpha l} = 4e^4(\delta_{ij}\delta_{kl} + \delta_{ik}\delta_{jl} + \delta_{il}\delta_{jk}) - 6e^4 \Delta_{ijkl}, \tag{3.14}$$

where $\Delta_{ijkl} = 1$ if $i = j = k = l$, otherwise $\Delta_{ijkl} = 0$.

Using Eq. (3.11) to evaluate the terms in Eq. (3.2), we have

$$N_\alpha = \frac{1}{e^2} \sum_\alpha e_{\alpha x} e_{\alpha x} = \frac{1}{e^2} \sum_\alpha e_{\alpha y} e_{\alpha y} = 6. \tag{3.15}$$

Substitution of the above equation into Eq. (3.10) leads to

$$f_\alpha(\mathbf{x} + \mathbf{e}_\alpha \Delta t, t + \Delta t) - f_\alpha(\mathbf{x}, t) = -\frac{1}{\tau}(f_\alpha - f_\alpha^{eq}) + \frac{\Delta t}{6e^2} e_{\alpha i} F_i, \tag{3.16}$$

which is the most common form of a lattice Boltzmann model used for simulating fluid flows.

3.4 Local Equilibrium Distribution Function

Determining a suitable local equilibrium function plays an essential role in the lattice Boltzmann method. It is this function that decides what flow equations are solved by means of the lattice Boltzmann equation (3.16). In order to apply the equation (3.16) for solution of the 2D shallow water equations (2.55) and (2.56), we derive a suitable local equilibrium function f_α^{eq} in this section.

According to the theory of the lattice gas automata, an equilibrium function is the Maxwell-Boltzmann equilibrium distribution function, which is often expanded as a Taylor series in macroscopic velocity to its second order [10, 38]. However, the use of such equilibrium function in the lattice Boltzmann equation can recover the Navier-Stokes equation only. This severely limits the capability of the method to solve flow equations. Thus an alternative and powerful way is to assume that an equilibrium function can be expressed as a power series in macroscopic velocity [39], i.e.,

$$f_\alpha^{eq} = A_\alpha + B_\alpha e_{\alpha i} u_i + C_\alpha e_{\alpha i} e_{\alpha j} u_i u_j + D_\alpha u_i u_i. \tag{3.17}$$

This turns out to be a general approach, which is successfully used for solution of various flow problems [40, 41], demonstrating its accuracy and suitability. Hence it is used. Since the equilibrium function has the same symmetry as the lattice (see Fig. 3.1), there must be

$$A_1 = A_3 = A_5 = A_7 = \bar{A}, \qquad A_2 = A_4 = A_6 = A_8 = \tilde{A}, \tag{3.18}$$

and similar expressions for B_α, C_α and D_α. Accordingly, it is convenient to write Eq. (3.17) in the following form,

$$f_\alpha^{eq} = \begin{cases} A_0 + D_0 u_i u_i, & \alpha = 0, \\ \bar{A} + \bar{B} e_{\alpha i} u_i + \bar{C} e_{\alpha i} e_{\alpha j} u_i u_j + \bar{D} u_i u_i, & \alpha = 1, 3, 5, 7, \\ \tilde{A} + \tilde{B} e_{\alpha i} u_i + \tilde{C} e_{\alpha i} e_{\alpha j} u_i u_j + \tilde{D} u_i u_i, & \alpha = 2, 4, 6, 8. \end{cases} \tag{3.19}$$

The coefficients such as A_0, $\bar{A}$ and $\tilde{A}$ can be determined based on the constraints on the equilibrium distribution function, i.e. it must obey the conservation relations such as mass and momentum conservations. For the shallow water equations, the local equilibrium distribution function (3.19) must satisfy the following three conditions,

$$\sum_\alpha f_\alpha^{eq}(\mathbf{x}, t) = h(\mathbf{x}, t), \tag{3.20}$$

$$\sum_\alpha e_{\alpha i} f_\alpha^{eq}(\mathbf{x}, t) = h(\mathbf{x}, t) u_i(\mathbf{x}, t), \tag{3.21}$$

$$\sum_\alpha e_{\alpha i} e_{\alpha j} f_\alpha^{eq}(\mathbf{x}, t) = \frac{1}{2} g h^2(\mathbf{x}, t) \delta_{ij} + h(\mathbf{x}, t) u_i(\mathbf{x}, t) u_j(\mathbf{x}, t). \tag{3.22}$$

Once the local equilibrium function (3.17) is determined under the above constraints, the calculation of the lattice Boltzmann equation (3.16) leads to the solution of the 2D shallow water equations (2.55) and (2.56) (The proof is given in Section 3.6).

Substituting Eq. (3.19) into Eq. (3.20) yields

$$\begin{aligned} &A_0 + D_0 u_i u_i + \\ &4\bar{A} + \sum_{i=1,3,5,7} \bar{B} e_{\alpha i} u_i + \sum_{i=1,3,5,7} \bar{C} e_{\alpha i} e_{\alpha j} u_i u_j + 4\bar{D} u_i u_i + \\ &4\tilde{A} + \sum_{i=2,4,6,8} \tilde{B} e_{\alpha i} u_i + \sum_{i=2,4,6,8} \tilde{C} e_{\alpha i} e_{\alpha j} u_i u_j + 4\tilde{D} u_i u_i = h. \end{aligned} \tag{3.23}$$

After evaluating the terms in the above equation with Eq. (3.11) and equating the coefficients of h and $u_i u_i$, respectively, we have

$$A_0 + 4\bar{A} + 4\tilde{A} = h, \tag{3.24}$$

and

$$D_0 + 2e^2 \bar{C} + 4e^2 \tilde{C} + 4\bar{D} + 4\tilde{D} = 0. \tag{3.25}$$

Inserting Eq. (3.19) to Eq. (3.21) leads to

$$\begin{aligned} &A_0 e_{\alpha i} + D_0 e_{\alpha i} u_j u_j + \\ &\sum_{\alpha=1,3,5,7} (\bar{A} e_{\alpha i} + \bar{B} e_{\alpha i} e_{\alpha j} u_j + \bar{C} e_{\alpha i} e_{\alpha j} e_{\alpha k} u_j u_k + \bar{D} e_{\alpha i} u_j u_j) + \\ &\sum_{\alpha=2,4,6,8} (\tilde{A} e_{\alpha i} + \tilde{B} e_{\alpha i} e_{\alpha j} u_j + \tilde{C} e_{\alpha i} e_{\alpha j} e_{\alpha k} u_j u_k + \tilde{D} e_{\alpha i} u_j u_j) = h u_i, \end{aligned} \tag{3.26}$$

from which we can obtain

$$2e^2 \bar{B} + 4e^2 \tilde{B} = h. \tag{3.27}$$

Substituting Eq. (3.19) into Eq. (3.22) results in

$$\begin{aligned} &\sum_{\alpha=1,3,5,7} (\bar{A} e_{\alpha i} e_{\alpha j} + \bar{B} e_{\alpha i} e_{\alpha j} e_{\alpha k} u_k + \bar{C} e_{\alpha i} e_{\alpha j} e_{\alpha k} e_{\alpha l} u_k u_l + \bar{D} e_{\alpha i} e_{\alpha j} u_k u_k) + \\ &\sum_{\alpha=2,4,6,8} (\tilde{A} e_{\alpha i} e_{\alpha j} + \tilde{B} e_{\alpha i} e_{\alpha j} e_{\alpha k} u_k + \tilde{C} e_{\alpha i} e_{\alpha j} e_{\alpha k} e_{\alpha l} u_k u_l + \tilde{D} e_{\alpha i} e_{\alpha j} u_k u_k) \\ &\qquad = \frac{1}{2} g h^2 \delta_{ij} + h u_i u_j. \end{aligned} \tag{3.28}$$

Use of Eq. (3.11) to simplify the above equation leads to

$$\begin{aligned} &2\bar{A} e^2 \delta_{ij} + 2\bar{C} e^4 u_i u_i + 2\bar{D} e^2 u_i u_i + 4\tilde{A} e^2 \delta_{ij} + \\ &8\tilde{C} e^4 u_i u_j + 4\tilde{C} e^4 u_i u_i + 4\tilde{D} e^2 u_i u_i = \frac{1}{2} g h^2 \delta_{ij} + h u_i u_j, \end{aligned} \tag{3.29}$$

which provides the following four relations,

$$2e^2\bar{A} + 4e^2\tilde{A} = \frac{1}{2}gh^2, \tag{3.30}$$

$$8e^4\tilde{C} = h, \tag{3.31}$$

$$2e^4\bar{C} = h, \tag{3.32}$$

$$2e^2\bar{D} + 4e^2\tilde{D} + 4e^4\tilde{C} = 0. \tag{3.33}$$

Combining of Eqs. (3.31) and (3.32) gives

$$\bar{C} = 4\tilde{C}. \tag{3.34}$$

From the symmetry of the lattice, based on Eq. (3.34), we have good reason to assume the three additional relations as follows,

$$\bar{A} = 4\tilde{A}, \tag{3.35}$$

$$\bar{B} = 4\tilde{B}, \tag{3.36}$$

$$\bar{D} = 4\tilde{D}. \tag{3.37}$$

Solution of Eqs. (3.24), (3.25), (3.27) and (3.30) - (3.37) results in

$$A_0 = h - \frac{5gh^2}{6e^2}, \qquad D_0 = -\frac{2h}{3e^2}, \tag{3.38}$$

$$\bar{A} = \frac{gh^2}{6e^2}, \qquad \bar{B} = \frac{h}{3e^2}, \qquad \bar{C} = \frac{h}{2e^4}, \qquad \bar{D} = -\frac{h}{6e^2}, \tag{3.39}$$

$$\tilde{A} = \frac{gh^2}{24e^2}, \qquad \tilde{B} = \frac{h}{12e^2}, \qquad \tilde{C} = \frac{h}{8e^4}, \qquad \tilde{D} = -\frac{h}{24e^2}. \tag{3.40}$$

Substitution of the above equations (3.38) - (3.40) into Eq. (3.19) leads to the following local equilibrium function,

$$f_\alpha^{eq} = \begin{cases} h - \frac{5gh^2}{6e^2} - \frac{2h}{3e^2}u_i u_i, & \alpha = 0, \\ \frac{gh^2}{6e^2} + \frac{h}{3e^2}e_{\alpha i}u_i + \frac{h}{2e^4}e_{\alpha i}e_{\alpha j}u_i u_j - \frac{h}{6e^2}u_i u_i, & \alpha = 1,3,5,7, \\ \frac{gh^2}{24e^2} + \frac{h}{12e^2}e_{\alpha i}u_i + \frac{h}{8e^4}e_{\alpha i}e_{\alpha j}u_i u_j - \frac{h}{24e^2}u_i u_i, & \alpha = 2,4,6,8, \end{cases} \tag{3.41}$$

which is used in the lattice Boltzmann equation (3.16) for solution of the shallow water equations (2.55) and (2.56).

3.5 Macroscopic Properties

The lattice Boltzmann equation (3.16) with the local equilibrium function (3.41) form a lattice Boltzmann model for shallow water flows (LABSWE) on the square lattices, which is described by Zhou [15]. The remaining task is how to determine the physical quantities, water depth h and velocity u_i, as the solution to the shallow water equations (2.55) and (2.56). For this purpose, we examine the macroscopic properties of the lattice Boltzmann equation (3.16). Taking the sum of the zeroth moment of the distribution function in Eq. (3.16) over the lattice velocities leads to

$$\sum_\alpha [f_\alpha(\mathbf{x}+\mathbf{e}_\alpha \Delta t, t+\Delta t) - f_\alpha(\mathbf{x}, t)] = -\frac{1}{\tau}\sum_\alpha (f_\alpha - f_\alpha^{eq}) + \frac{\Delta t}{6e^2}\sum_\alpha e_{\alpha i} F_i. \quad (3.42)$$

Notice $\sum_\alpha e_{\alpha i} F_i = 0$, Eq. (3.42) becomes

$$\sum_\alpha [f_\alpha(\mathbf{x}+\mathbf{e}_\alpha \Delta t, t+\Delta t) - f_\alpha(\mathbf{x}, t)] = -\frac{1}{\tau}\sum_\alpha (f_\alpha - f_\alpha^{eq}). \quad (3.43)$$

In the lattice Boltzmann method, we have an explicit constraint to preserve conservative property, i.e. the cumulative mass and momentum which are the corresponding summations of the microdynamic mass and momentum are conserved. The mass conservation requires the following identity,

$$\sum_\alpha f_\alpha(\mathbf{x}+\mathbf{e}_\alpha \Delta t, t+\Delta t) \equiv \sum_\alpha f_\alpha(\mathbf{x}, t), \quad (3.44)$$

which is the continuity equation with microdynamic variables.

Substitution of the above equation into Eq. (3.43) leads to

$$\sum_\alpha f_\alpha(\mathbf{x}, t) = \sum_\alpha f_\alpha^{eq}(\mathbf{x}, t). \quad (3.45)$$

With reference to Eq. (3.20), the above expression in fact results in the definition for the physical quantity, water depth h, as

$$h(\mathbf{x}, t) = \sum_\alpha f_\alpha(\mathbf{x}, t). \quad (3.46)$$

Now, in order to find the expression for the velocity, we take the sum of the first moment of the distribution function in Eq. (3.16) over the lattice velocities,

$$\sum_\alpha e_{\alpha i}[f_\alpha(\mathbf{x}+\mathbf{e}_\alpha \Delta t, t+\Delta t) - f_\alpha(\mathbf{x}, t)] = -\frac{1}{\tau}\sum_\alpha e_{\alpha i}(f_\alpha - f_\alpha^{eq}) + \frac{\Delta t}{6e^2}\sum_\alpha e_{\alpha i} e_{\alpha j} F_j, \quad (3.47)$$

which can be simplified with Eq. (3.13) and rearranged as

$$\sum_\alpha e_{\alpha i}[f_\alpha(\mathbf{x}+\mathbf{e}_\alpha \Delta t, t+\Delta t) - f_\alpha(\mathbf{x},t)] = F_i \Delta t - \frac{1}{\tau}\sum_\alpha e_{\alpha i}(f_\alpha - f_\alpha^{eq}). \quad (3.48)$$

which reflects the evolution of cumulative momentum in the distribution function. Again, the momentum conservation requires the following identity,

$$\sum_\alpha e_{\alpha i}[f_\alpha(\mathbf{x}+\mathbf{e}_\alpha \Delta t, t+\Delta t) - f_\alpha(\mathbf{x},t)] \equiv F_i \Delta t. \quad (3.49)$$

which is the momentum equation with microdynamic variables, representing the Newton's second law.

Substitution of Eq. (3.49) into Eq. (3.48) provides

$$\sum_\alpha e_{\alpha i} f_\alpha(\mathbf{x},t) = \sum_\alpha e_{\alpha i} f_\alpha^{eq}(\mathbf{x},t). \quad (3.50)$$

The use of Eq. (3.21) in the above equation leads to the definition for another physical variable, velocity u_i, as

$$u_i(\mathbf{x},t) = \frac{1}{h(\mathbf{x},t)}\sum_\alpha e_{\alpha i} f_\alpha(\mathbf{x},t). \quad (3.51)$$

As can be seen from Eqs. (3.20), (3.21), (3.46) and (3.51), it seems that there are redundant definitions for the physical variables h and u_i. However, a careful examination indicates that this is just an important feature which is peculiar to the lattice Boltzmann method. First of all, the local equilibrium function f_α^{eq} given by (3.41) satisfy Eqs. (3.20) and (3.21). Secondly, the distribution function f_α relaxes to its local equilibrium function f_α^{eq} via the lattice Boltzmann equation (3.16). Finally, the physical variables determined from Eqs. (3.46) and (3.51) will guarantee that both Eqs. (3.45) and (3.50) hold true, hence preserving the two identities (3.44) and (3.49) during the numerical procedure. This makes the lattice Boltzmann method very elegant and effectively explains why the method is accurate and conservative.

3.6 Recovery of the Shallow Water Equations

In order to prove that the depth and the velocities calculated from Eqs. (3.46) and (3.51) are the solution to the shallow water equations, we perform the Chapman-Enskog expansion to the lattice Boltzmann equation (3.16) that recovers the macroscopic equations (2.55) and (2.56).

Assuming Δt is small and is equal to ε,

$$\Delta t = \varepsilon, \quad (3.52)$$

we have the equation (3.16) expressed as

$$f_\alpha(\mathbf{x}+\mathbf{e}_\alpha\varepsilon, t+\varepsilon) - f_\alpha(\mathbf{x}, t) = -\frac{1}{\tau}(f_\alpha - f_\alpha^{eq}) + \frac{\varepsilon}{6e^2} e_{\alpha j} F_j. \tag{3.53}$$

Taking a Taylor expansion to the first term on the left-hand side of the above equation in time and space around point $(\mathbf{x}, t)$ leads to

$$\varepsilon(\frac{\partial}{\partial t} + e_{\alpha j}\frac{\partial}{\partial x_j}) f_\alpha + \frac{1}{2}\varepsilon^2(\frac{\partial}{\partial t} + e_{\alpha j}\frac{\partial}{\partial x_j})^2 f_\alpha + \mathcal{O}(\varepsilon^2) = -\frac{1}{\tau}(f_\alpha - f_\alpha^{(0)}) + \frac{\varepsilon}{6e^2} e_{\alpha j} F_j. \tag{3.54}$$

We can also expand f_α around $f_\alpha^{(0)}$,

$$f_\alpha = f_\alpha^{(0)} + \varepsilon f_\alpha^{(1)} + \varepsilon^2 f_\alpha^{(2)} + \mathcal{O}(\varepsilon^2), \tag{3.55}$$

where $f_\alpha^{(0)} = f_\alpha^{eq}$.

The equation (3.54) to order ε is

$$(\frac{\partial}{\partial t} + e_{\alpha j}\frac{\partial}{\partial x_j}) f_\alpha^{(0)} = -\frac{1}{\tau} f_\alpha^{(1)} + \frac{1}{6e^2} e_{\alpha j} F_j \tag{3.56}$$

and to order ε^2 is

$$(\frac{\partial}{\partial t} + e_{\alpha j}\frac{\partial}{\partial x_j}) f_\alpha^{(1)} + \frac{1}{2}(\frac{\partial}{\partial t} + e_{\alpha j}\frac{\partial}{\partial x_j})^2 f_\alpha^{(0)} = -\frac{1}{\tau} f_\alpha^{(2)}. \tag{3.57}$$

Substitution of Eq. (3.56) into Eq. (3.57), after rearranged, gives

$$(1 - \frac{1}{2\tau})(\frac{\partial}{\partial t} + e_{\alpha j}\frac{\partial}{\partial x_j}) f_\alpha^{(1)} = -\frac{1}{\tau} f_\alpha^{(2)} - \frac{1}{2}(\frac{\partial}{\partial t} + e_{\alpha j}\frac{\partial}{\partial x_j})(\frac{1}{6e^2} e_{\alpha k} F_k). \tag{3.58}$$

Taking $\sum$ [(3.56) + $\varepsilon\times$ (3.58)] about α provides

$$\frac{\partial}{\partial t}(\sum_\alpha f_\alpha^{(0)}) + \frac{\partial}{\partial x_j}(\sum_\alpha e_{\alpha j} f_\alpha^{(0)}) = -\varepsilon \frac{1}{12e^2}\frac{\partial}{\partial x_j}(\sum_\alpha e_{\alpha j} e_{\alpha k} F_k). \tag{3.59}$$

If the first-order accuracy for the force term is applied, evaluation of the other terms in the above equation using Eqs. (3.11) and (3.41) results in

$$\frac{\partial h}{\partial t} + \frac{\partial (h u_j)}{\partial x_j} = 0, \tag{3.60}$$

which is the continuity equation (2.55) for shallow water flows.

From $\sum e_{\alpha i}$[(3.56) + $\varepsilon\times$ (3.58)] about α, we have

$$\frac{\partial}{\partial t}(\sum_\alpha e_{\alpha i} f_\alpha^{(0)}) + \frac{\partial}{\partial x_j}(\sum_\alpha e_{\alpha i} e_{\alpha j} f_\alpha^{(0)}) + \varepsilon(1 - \frac{1}{2\tau})\frac{\partial}{\partial x_j}(\sum_\alpha e_{\alpha i} e_{\alpha j} f_\alpha^{(1)})$$
$$= F_j \delta_{ij} - \varepsilon\frac{1}{2}\sum_\alpha e_{\alpha i}(\frac{\partial}{\partial t} + e_{\alpha j}\frac{\partial}{\partial x_j})(\frac{1}{6e^2} e_{\alpha j} F_j). \tag{3.61}$$

Again, if the first-order accuracy for the force term is used, after the other terms is simplified with Eqs. (3.11) and (3.41), the above equation becomes

$$\frac{\partial(hu_i)}{\partial t} + \frac{\partial(hu_iu_j)}{\partial x_j} = -g\frac{\partial}{\partial x_i}(\frac{h^2}{2}) - \frac{\partial}{\partial x_j}\Lambda_{ij} + F_i, \tag{3.62}$$

with

$$\Lambda_{ij} = \frac{\varepsilon}{2\tau}(2\tau - 1)\sum_\alpha e_{\alpha i}e_{\alpha j}f_\alpha^{(1)}. \tag{3.63}$$

With reference to Eq. (3.56), using Eqs. (3.11) and (3.41), after some algebra, we obtain

$$\Lambda_{ij} \approx -\nu\left[\frac{\partial(hu_i)}{\partial x_j} + \frac{\partial(hu_j)}{\partial x_i}\right]. \tag{3.64}$$

Substitution of Eq. (3.64) into Eq. (3.62) leads to

$$\frac{\partial(hu_i)}{\partial t} + \frac{\partial(hu_iu_j)}{\partial x_j} = -g\frac{\partial}{\partial x_i}(\frac{h^2}{2}) + \nu\frac{\partial^2(hu_i)}{\partial x_j\partial x_j} + F_i, \tag{3.65}$$

with the kinematic viscosity ν defined by

$$\nu = \frac{e^2\Delta t}{6}(2\tau - 1) \tag{3.66}$$

and the force term F_i expressed as

$$F_i = -gh\frac{\partial z_b}{\partial x_i} + \frac{\tau_{wi}}{\rho} - \frac{\tau_{bi}}{\rho} + E_i. \tag{3.67}$$

The equation (3.65) is just the momentum equation (2.56) for the shallow water equations.

It should be pointed that the above proof shows that the lattice Boltzmann equation (3.16) is only first-order accurate for the recovered macroscopic continuity and momentum equations. We can prove that the use of a suitable form for the force term can make Eq. (3.16) second-order accurate for the recovered macroscopic continuity and momentum equations (the detail is given in Section 4).

3.7 Stability Conditions

The lattice Boltzmann equation (3.16) is a discrete form of a numerical method. It may suffer from a numerical instability like any other numerical methods. Theoretically, the stability conditions are not generally known for the method. In practice, a lot of computations have shown that the method is often stable if some basic conditions are satisfied. They are described now.

First of all, if a solution from the lattice Boltzmann equation (3.16) represents a physical water flow, there must be diffusion phenomena. This implies that the kinematic viscosity ν should be positive [42], i.e. from Eq. (3.66) we must have

$$\nu = \frac{e^2 \Delta t}{6}(2\tau - 1) > 0. \tag{3.68}$$

Thus an apparent constraint on the relaxation time is

$$\tau > \frac{1}{2}. \tag{3.69}$$

Secondly, the magnitude of the resultant velocity is smaller than the speed calculated with the lattice size divided by the time step,

$$\frac{u_j u_j}{e^2} < 1, \tag{3.70}$$

and so is the celerity,

$$\frac{gh}{e^2} < 1. \tag{3.71}$$

Finally, since the lattice Boltzmann method is limited to low speed flows, this suggests that the current lattice Boltzmann method is suitable for subcritical shallow water flows, which brings the final constraint on the method as

$$\frac{u_j u_j}{gh} < 1, \tag{3.72}$$

because the term on the left hand side of the above expression is equivalent to the definition of the Froude number,

$$F_r = \frac{\sqrt{u_j u_j}}{\sqrt{gh}}, \tag{3.73}$$

in hydraulics. The Froude number can be used to decide a flow state for flows with free surface, i.e. it is a subcritical flow if $F_r < 1$, a critical flow if $F_r = 1$, and a supercritical flow if $F_r > 1$. Consequently, the condition (3.72) indicates that the lattice Boltzmann method is suitable for subcritical flows. Since such flows are the most scenarios in coastal areas, estuaries and harbours, the condition (3.72) is normally satisfied.

It should be pointed out that the first three conditions (3.69) - (3.71) can be easily satisfied by using suitable values for the relaxation time, the lattice size and the time step. Thus, in general, the lattice Boltzmann method is stable as long as these four stability conditions are satisfied.

3.8 Relation to Continuum Boltzmann Equation

Historically, the lattice Boltzmann equation (3.16) is evolved from the lattice gas automata. The researches [38, 43] have shown that the lattice Boltzmann equation can also be obtained from the continuum Boltzmann equation. This may be explained as follows.

A simple kinetic model is the Boltzmann BGK equation [14],

$$\frac{\partial f}{\partial t} + \mathbf{e} \cdot \nabla f = -\frac{1}{\lambda}(f - f^{eq}). \tag{3.74}$$

where $f = f(\mathbf{x}, \mathbf{e}, t)$ is the single-particle distribution in continuum phase space $(\mathbf{x}, \mathbf{e})$, $\mathbf{e}$ is the particle velocity, $\nabla = \imath\frac{\partial}{\partial x} + \jmath\frac{\partial}{\partial y}$ is the gradient operator and λ is a relaxation time, and f^{eq} is the Maxwell-Boltzmann equilibrium distribution function defined by

$$f^{eq} = \frac{\rho}{(2\pi/3)^{D/2}} \exp\left[-\frac{3}{2}(\mathbf{e} - \mathbf{V})^2\right], \tag{3.75}$$

in which D is the spatial dimension; particle velocity $\mathbf{e}$ and fluid velocity $\mathbf{V}$ are nomralised by $\sqrt{3RT}$ (here R is the ideal gas constant and T is temperature), which gives a sound speed of $U_s = 1/\sqrt{3}$ [10]. The fluid density and velocity are calculated in terms of the distribution function,

$$\rho = \int f d\mathbf{e}, \qquad \rho\mathbf{V} = \int \mathbf{e} f d\mathbf{e}. \tag{3.76}$$

If the fluid velocity $\mathbf{V}$ is small compared with the sound speed, the equilibrium distribution function given by Eq. (3.75) can be expanded in the following form up to the second-order accuracy [44],

$$f^{eq} = \frac{\rho}{(2\pi/3)^{D/2}} \exp\left(-\frac{3}{2}\mathbf{e}^2\right)\left[1 + 3(\mathbf{e}\cdot\mathbf{V}) + \frac{9}{2}(\mathbf{e}\cdot\mathbf{V})^2 - \frac{3}{2}\mathbf{V}\cdot\mathbf{V}\right]. \tag{3.77}$$

In order to develop a discrete model, a limited number of particle velocities instead of the whole, i.e. $\mathbf{e}_\alpha$ $(\alpha = 1, ..., M)$ are used and hence their distribution functions at these velocities are

$$f_\alpha(\mathbf{x}, t) = f(\mathbf{x}, \mathbf{e}_\alpha, t), \qquad f_\alpha^{eq}(\mathbf{x}, t) = f^{eq}(\mathbf{x}, \mathbf{e}_\alpha, t), \tag{3.78}$$

which also satisfies Eq. (3.74),

$$\frac{\partial f_\alpha}{\partial t} + \mathbf{e}_\alpha \cdot \nabla f_\alpha = -\frac{1}{\lambda}(f_\alpha - f_\alpha^{eq}). \tag{3.79}$$

Since in the limited discrete space and time, the particles are allowed to move along their velocities, the above equation describes the change in the distribution function along its moving direction from a lattice point to its neighboring lattice point, which is effectively the lattice Boltzmann equation represented in the Lagrangian means. The left hand side of Eq. (3.79) is the Lagrangian time derivative and can be written as

$$\frac{\partial f_\alpha}{\partial t} + \mathbf{e}_\alpha \cdot \nabla f_\alpha = \frac{D f_\alpha}{Dt}, \tag{3.80}$$

which has the standard discretized form,

$$\frac{Df_\alpha}{Dt} = \frac{f_\alpha(\mathbf{x}+\mathbf{e}_\alpha \Delta t, t+\Delta t) - f_\alpha(\mathbf{x},t)}{\Delta t}. \tag{3.81}$$

Substitution of the above expression into Eq. (3.79) results in the standard lattice Boltzmann equation,

$$f_\alpha(\mathbf{x}+\mathbf{e}_\alpha \Delta t, t+\Delta t) - f_\alpha(\mathbf{x},t) = -\frac{1}{\tau}(f_\alpha - f_\alpha^{eq}), \tag{3.82}$$

with $\tau = \Delta t/\lambda$. Strictly speaking, τ should be called a single dimensionless relaxation time.

3.9 Discussions

3.9.1 Two Variants of the LBE

The lattice Boltzmann method is developed and developing rapidly in recent years. Now, there are two main variants of the standard lattice Boltzmann equation that are used to solve flow problems. One is the equation (3.79) and another is so-called a differential LBE, which is often derived from Eq. (3.16) via the Taylor series, For example, the use of the following Taylor series,

$$f_\alpha(\mathbf{x}+\mathbf{e}_\alpha \Delta t, t+\Delta t) = f_\alpha(\mathbf{x},t)+\Delta t\left(\frac{\partial}{\partial t} + e_{\alpha j}\frac{\partial}{\partial x_j}\right) f_\alpha(\mathbf{x},t)+\mathcal{O}(\Delta t^2), \tag{3.83}$$

reverts the standard lattice Boltzmann equation (3.16) to a differential LBE,

$$\frac{\partial f_\alpha}{\partial t} + e_{\alpha j}\frac{\partial f_\alpha}{\partial x_j} = -\frac{1}{\tau \Delta t}(f_\alpha - f_\alpha^{eq}) + \frac{1}{6e^2}e_{\alpha i}F_i. \tag{3.84}$$

As a result, either Eq. (3.79) or Eq. (3.84) can be solved with a conventional solution method such as finite difference method, finite volume method [45]. The main advantage is that these equations are also suitable for use with co-ordinate transformation. This is useful for flows with complicated geometries. However, the apparent weakness is that they lose the locality in the kinetic approach and simple arithmetic calculations that are the basic feature of the lattice Boltzmann method, which undoubtedly prevents the methods from solving complex flows, for instance, flows through porous media.

3.9.2 Solution Strategies

In the literature, there are many solution strategies suggested for use with the standard lattice Boltzmann equation (3.16), differential LBE (3.79) and (3.84). They can be used with uniform and non-uniform lattices. Also, a multi-block method and composite grids are introduced for the application in lattice Boltzmann method [46, 47]. Although these provide some capabilities for some flows, the use of Eq. (3.16) with uniform lattices is a simple and universal method which is suitable for any flow problems.

3.9.3 Choice of the LBE

As described in the above two sections, the differential LBE involves solution method similar to convectional numerical method with approximation to derivatives and its procedure is more complicated compared with that for the standard lattice Boltzmann equation (3.16). The use of either non-uniform lattices or composite grids will limit the capabilities of the LBM in many complex situations. Therefore, the standard lattice Boltzmann equation (3.16) together with the uniform lattices is still the most popular version of the lattice Boltzmann equation in use today, representing the mainstream of the method. It offers the maximum capabilities and advantages as indicated in Section 1.3. This is the reason why it is chosen as the basis for the method presented in this book.

3.10 Closure

In the lattice Boltzmann method, the central quantity is the particle distribution function f_α which is governed by the lattice Boltzmann equation (3.16). Its evolution is carried out on the square lattices in the computational domain. The physical variables, water depth and velocities, are calculated in terms of the distribution function via Eqs. (3.46) and (3.51). It has shown that the depth and the velocities determined in this way satisfy the shallow water equations. The four conditions for stability are described, which are very useful in practical computations. Apparently, the lattice Boltzmann method is a numerical technique for an indirect solution of flow equations through a microscopic approach to a macroscopic phenomenon. This bridges the gap between discrete microscopic and continuum macroscopic phenomena, providing a very powerful tool for modeling a wide variety of complex fluid flows.

4

Force Terms

4.1 Introduction

In Chapter 3, we have described a lattice Boltzmann model for shallow water equations (LABSWE). Although a force term is included in the model, how to evaluate it is not discussed there. In fact, most studies reported in the literature are restricted to the standard lattice Boltzmann equation without a force term. This severely limits the capability of the method to simulate fluid flows, because realistic flow problems usually involve different forces such as gravity, Coriolis and bed slope. Recently, a centred scheme for accurate evaluation of force terms in the lattice Boltzmann equation (3.16) is introduced by the author [18], indicating that a correct determination of a force term is essential for an accurate solution to flow equations with force terms. Thus, the centred scheme and its features are described and discussed here.

4.2 Motivation

Since accurate predictions of flows involving an external force require a proper treatment for force terms, the study of a force term in the lattice Boltzmann method begins to catch attention. Martys el al. [48] represented a force term based on a Hermite expansion. But they did not give a numerical example. Also, the scheme is complicated. Buick and Greated [49] described a composite method to consider the gravity in a lattice Boltzmann model. Moreover, they reviewed another three methods to include gravity in the lattice Boltzmann equation: 1. combining the gravity term and pressure tensor which is suitable for flows with negligible change in density; 2. calculating the equilibrium distribution with an altered velocity, which satisfies the Navier-Stokes equation with unnecessary terms instead of the physical velocity, hence introducing further approximation in the recovered macroscopic equations; and 3. adding gravity as an additional term to the collision term in the lattice Boltzmann equation, which lacks a physical interpretation. Since the composite method of Buick and Greated is a combination of the methods 2 and 3, it has the same weakness.

Recently, Zhou [15, 17] has successfully demonstrated that direct incorporation of force terms such as wind shear stress and bed slope into the streaming step in the lattice Boltzmann equation is a simple and general method, which represents the underlying physics, producing accurate solutions to many flows. A force term is simply evaluated in a straightforward way. The advantage is that it retains the same definition as the standard lattice Boltzmann equation and any additional force terms can naturally be taken into account. The method preserves the efficiency of the lattice Boltzmann method; hence it is preferred in practical computation. However, after carrying out a few further numerical tests, Zhou has noticed that such a straightforward evaluation of a force term in the lattice Boltzmann equation cannot generate an accurate solution to some flow problems involving a force. This inspires the author to develop a centred scheme [18], which is described in the following section.

4.3 Centred Scheme

Mathematically, it seems that it is straightforward to calculate a force term from Eq. (3.67) for the lattice Boltzmann equation (3.16). However, the numerical tests indicate that a solution in this way via Eq. (3.16) is not accurate for some flow problems, suggesting that the force term is not correctly determined. Evidently, how to determine the force term in the lattice Boltzmann equation (3.16) is essential for accurate numerical solutions. Here we introduce three schemes for representing F_i as follows.

- Basic scheme: this is an apparent way - the force term is evaluated at the lattice point with
$$F_i = F_i(\mathbf{x}, t). \tag{4.1}$$
- Second-order scheme: the force term takes the averaged value of the two values at the lattice point and its neighboring lattice point respectively,
$$F_i = \frac{1}{2}[F_i(\mathbf{x}, t) + F_i(\mathbf{x} + \mathbf{e}_\alpha \Delta t, t + \Delta t)], \tag{4.2}$$
which may take the form of
$$F_i = \frac{1}{2}[F_i(\mathbf{x}, t) + F_i(\mathbf{x} + \mathbf{e}_\alpha \Delta t, t)] \tag{4.3}$$
for the purpose of parallel process.
- Centred scheme: the force term is evaluated at the mid-point between the lattice point and its neighboring lattice point as
$$F_i = F_i(\mathbf{x} + \frac{1}{2}\mathbf{e}_\alpha \Delta t, t + \frac{1}{2}\Delta t) \tag{4.4}$$

which can also be expressed in the following semi-implicit form,

$$F_i = F_i(\mathbf{x} + \frac{1}{2}\mathbf{e}_\alpha \Delta t, t), \tag{4.5}$$

suitable for parallel process.

Now, we apply the Chapman-Enskog procedure to analyze the features of these schemes. If the centred scheme is used for the force term, assuming Δt is small and equal to ε, i.e. $\Delta t = \varepsilon$, we write the equation (3.16) as

$$\begin{aligned} f_\alpha(\mathbf{x} + \mathbf{e}_\alpha \varepsilon, t + \varepsilon) - f_\alpha(\mathbf{x}, t) = &-\frac{1}{\tau}[f_\alpha(\mathbf{x}, t) - f_\alpha^{eq}(\mathbf{x}, t)] \\ &+ \frac{\varepsilon}{6e^2} e_{\alpha i} F_i(\mathbf{x} + \frac{1}{2}\mathbf{e}_\alpha \varepsilon, t + \frac{1}{2}\varepsilon). \end{aligned} \tag{4.6}$$

Taking a Taylor expansion to the first term on the left-hand side of the above equation in time and space around point $(\mathbf{x}, t)$,

$$\begin{aligned} f_\alpha(\mathbf{x} + \mathbf{e}_\alpha \varepsilon, t + \varepsilon) = &f_\alpha(\mathbf{x}, t) + \varepsilon(\frac{\partial}{\partial t} + e_{\alpha j}\frac{\partial}{\partial x_j}) f_\alpha \\ &+ \frac{1}{2}\varepsilon^2(\frac{\partial}{\partial t} + e_{\alpha j}\frac{\partial}{\partial x_j})^2 f_\alpha + O(\varepsilon^2), \end{aligned} \tag{4.7}$$

and also to the force term on the right-hand side,

$$F_i(\mathbf{x} + \frac{1}{2}\mathbf{e}_\alpha \varepsilon, t + \frac{1}{2}\varepsilon) = F_i(\mathbf{x}, t) + \frac{1}{2}\varepsilon(\frac{\partial}{\partial t} + e_{\alpha j}\frac{\partial}{\partial x_j}) F_i(\mathbf{x}, t) + O(\varepsilon). \tag{4.8}$$

Substitution of Eqs. (4.7) and (4.8) into Eq. (4.6) leads to

$$\begin{aligned} \varepsilon(\frac{\partial}{\partial t} + e_{\alpha j}\frac{\partial}{\partial x_j}) f_\alpha + \frac{1}{2}\varepsilon^2(\frac{\partial}{\partial t} + e_{\alpha j}\frac{\partial}{\partial x_j})^2 f_\alpha = &-\frac{1}{\tau}(f_\alpha - f_\alpha^{(0)}) + \frac{\varepsilon}{6e^2} e_{\alpha i} F_i \\ &+ \frac{\varepsilon^2}{12e^2}(\frac{\partial}{\partial t} + e_{\alpha j}\frac{\partial}{\partial x_j}) e_{\alpha i} F_i + O(\varepsilon^2). \end{aligned} \tag{4.9}$$

By expanding f_α around $f_\alpha^{(0)}$,

$$f_\alpha = f_\alpha^{(0)} + \varepsilon f_\alpha^{(1)} + \varepsilon^2 f_\alpha^{(2)} + O(\varepsilon^2), \tag{4.10}$$

where $f_\alpha^{(0)} = f_\alpha^{eq}$, we obtain the equation (4.9) to order ε as

$$(\frac{\partial}{\partial t} + e_{\alpha j}\frac{\partial}{\partial x_j}) f_\alpha^{(0)} = -\frac{1}{\tau} f_\alpha^{(1)} + \frac{1}{6e^2} e_{\alpha i} F_i, \tag{4.11}$$

and to order ε^2 as

$$\begin{aligned} (\frac{\partial}{\partial t} + e_{\alpha j}\frac{\partial}{\partial x_j}) f_\alpha^{(1)} + \frac{1}{2}(\frac{\partial}{\partial t} + e_{\alpha j}\frac{\partial}{\partial x_j})^2 f_\alpha^{(0)} = &-\frac{1}{\tau} f_\alpha^{(2)} \\ &+ \frac{1}{12e^2}(\frac{\partial}{\partial t} + e_{\alpha j}\frac{\partial}{\partial x_j}) e_{\alpha i} F_i. \end{aligned} \tag{4.12}$$

Substitution of Eq. (4.11) into Eq. (4.12) gives

$$(1-\frac{1}{2\tau})(\frac{\partial}{\partial t}+e_{\alpha j}\frac{\partial}{\partial x_j})f_\alpha^{(1)}=-\frac{1}{\tau}f_\alpha^{(2)}. \tag{4.13}$$

Taking $\sum$ [(4.11) + $\varepsilon\times$ (4.13)] about α provides

$$\frac{\partial}{\partial t}(\sum_\alpha f_\alpha^{(0)})+\frac{\partial}{\partial x_j}(\sum_\alpha e_{\alpha j}f_\alpha^{(0)})=0. \tag{4.14}$$

Evaluation of the terms in the above equation using Eq. (3.41) results in the second-order accurate continuity equation (2.55).

Taking $\sum e_{\alpha i}$[(4.11) + $\varepsilon\times$ (4.13)] about α yields

$$\frac{\partial}{\partial t}(\sum_\alpha e_{\alpha i}f_\alpha^{(0)})+\frac{\partial}{\partial x_j}(\sum_\alpha e_{\alpha i}e_{\alpha j}f_\alpha^{(0)}) + \varepsilon(1-\frac{1}{2\tau})\frac{\partial}{\partial x_j}(\sum_\alpha e_{\alpha i}e_{\alpha j}f_\alpha^{(1)})=F_i. \tag{4.15}$$

After the terms are simplified with Eq. (3.41), the above equation becomes

$$\frac{\partial(hu_i)}{\partial t}+\frac{\partial(hu_iu_j)}{\partial x_j}=-g\frac{\partial}{\partial x_i}(\frac{h^2}{2})-\frac{\partial}{\partial x_j}\Lambda_{ij}+F_i, \tag{4.16}$$

where

$$\Lambda_{ij}=\frac{\varepsilon}{2\tau}(2\tau-1)\sum_\alpha e_{\alpha i}e_{\alpha j}f_\alpha^{(1)}. \tag{4.17}$$

With reference to Eq. (4.11), using Eq. (3.41), after some algebra, we obtain

$$\Lambda_{ij}\approx-\frac{e^2\varepsilon}{6}(2\tau-1)\left[\frac{\partial(hu_i)}{\partial x_j}+\frac{\partial(hu_j)}{\partial x_i}\right]. \tag{4.18}$$

Substitution of Eq. (4.18) into Eq. (4.16) results in the momentum equation (2.56), which is second-order accurate.

Therefore, the lattice Boltzmann equation with the centred scheme for the force term can generate the second-order accurate macroscopic continuity and momentum equations in time and space.

In an analogous manner, we can prove that the use of the second-order scheme (4.2) for the force term in Eq. (3.16) also results in second-order accurate macroscopic equations in time and space. However, the use of the basic scheme (4.1) provides only first-order accurate macroscopic equations in time and space as shown in Section 3.6.

The problem is which scheme should be used in the lattice Boltzmann equation (3.16) for an accurate solution. As indicated in Section 3.2, the lattice Boltzmann equation consists of a streaming and a collision steps, implying inclusion of a force term can be achieved by adding it either in the collision step or in the streaming step. Although these two ways lead to the same lattice Boltzmann equation (3.16), the recent research [18] reveals that Eq. (3.16) is based on Eqs. (3.1) and (3.3) with incorporation of the force term into the streaming step is preferred due to the fact that any suitable form for a force term within the streaming path may be used. This undoubtedly provides an additional freedom to determine the force term, i.e. F_i can theoretically be determined with either the basic scheme, or the second-order scheme, or the centred scheme. From the point of view of mathematics, both the centred scheme and the second-order scheme are the apparent choice because both can generate second-order accurate macroscopic equations. From the point of view of physics, which also truly represents the underlying physics is the correct choice and is expected to produce accurate solutions. In general, the particles' collision occurs within much shorter time compared with their streaming. Thus, it is physically reasonable that the force has a dominant effect on the particles during the streaming step, i.e. incorporation of the force term into the streaming step reflects the underlying physics. This suggests that the averaged force acting on the particles during streaming can be best represented only with the centred scheme; hence the centred scheme is the correct choice for determining the force term in the lattice Boltzmann equation (3.16).

4.4 Feature of the Centred Scheme

A correct numerical method can generate accurate predictions to both stationary and flow problems. Since there is an exact solution for a stationary case, whether or not a numerical method can produce an accurate solution to such case is in fact a basic criterion, which is also a fundamental rule for developing a new numerical method. Therefore, we introduce the following necessary property for an accurate numerical scheme.

Definition 4.1. *A numerical scheme is said to satisfy the necessary property ($\mathcal{N}$-property) provided that it can replicate the exact solution to a stationary case $u_i \equiv 0$ in which there is a non-zero force or source term.*

For example, there is a non-vanishing force term in the lattice Boltzmann equation (3.16) for shallow water equations with $u_i \equiv 0$ for still water above an uneven bed topography shown in Fig. 4.1, i.e.

$$F_i = -gh\frac{\partial z_b}{\partial x_i}. \tag{4.19}$$

How this term is exactly canceled such that the lattice Boltzmann equation can produce an accurate solution to such a stationary case ($u_i \equiv 0$) is obviously the first criterion that should be used to justify a numerical scheme for its correctness.

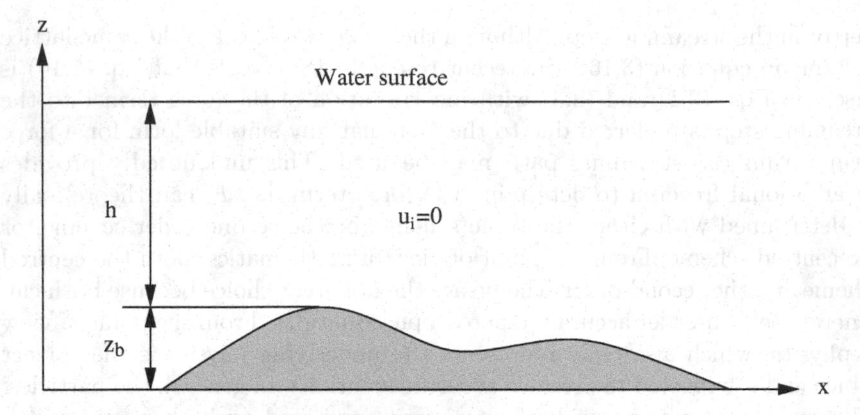

Fig. 4.1. Still water above an uneven bed.

Proposition 4.2. *The centred scheme proposed in Section 4.3 satisfies the $\mathcal{N}$-property at second-order accuracy.*

Proof. First of all, we take a Taylor expansion to the first term on the left-hand side of Eq. (3.16) without an approximation as

$$f_\alpha(\mathbf{x} + \mathbf{e}_\alpha \varepsilon, t + \varepsilon) = f_\alpha(\mathbf{x}, t) + \varepsilon(\frac{\partial}{\partial t} + e_{\alpha j}\frac{\partial}{\partial x_j}) f_\alpha(\mathbf{x} + \beta \mathbf{e}_\alpha \varepsilon, t + \beta\varepsilon), \quad (4.20)$$

where

$$0 < \beta < 1. \quad (4.21)$$

Substitution of the above equation into Eq. (3.16) results in

$$\varepsilon(\frac{\partial}{\partial t} + e_{\alpha j}\frac{\partial}{\partial x_j}) f_\alpha(\mathbf{x} + \beta \mathbf{e}_\alpha \varepsilon, t + \beta\varepsilon) = -\frac{1}{\tau}(f_\alpha - f_\alpha^{eq}) + \frac{\varepsilon}{6e^2} e_{\alpha i} F_i(\mathbf{x} + \frac{1}{2}\mathbf{e}_\alpha \varepsilon, t + \frac{1}{2}\varepsilon). \quad (4.22)$$

Secondly, after taking $\sum e_{\alpha i} \cdot$ (4.22), we have

$$\frac{\partial}{\partial t}\sum_\alpha e_{\alpha i} f_\alpha(\mathbf{x} + \beta \mathbf{e}_\alpha \varepsilon, t + \beta\varepsilon) + \frac{\partial}{\partial x_j}\sum_\alpha e_{\alpha i} e_{\alpha j} f_\alpha(\mathbf{x} + \beta \mathbf{e}_\alpha \varepsilon, t + \beta\varepsilon) = F_i(\mathbf{x} + \frac{1}{2}\mathbf{e}_\alpha \varepsilon, t + \frac{1}{2}\varepsilon). \quad (4.23)$$

According to the lattice Boltzmann method, e.g. see Eq. (3.50), we have

$$\sum_\alpha e_{\alpha i} f_\alpha(\mathbf{x} + \beta \mathbf{e}_\alpha \varepsilon, t + \beta\varepsilon) = \sum_\alpha e_{\alpha i} f_\alpha^{(eq)}(\mathbf{x} + \beta \mathbf{e}_\alpha \varepsilon, t + \beta\varepsilon). \quad (4.24)$$

With reference to Eq. (4.10), we obtain

$$\sum_\alpha e_{\alpha i} e_{\alpha j} f_\alpha(\mathbf{x} + \beta \mathbf{e}_\alpha \varepsilon, t + \beta\varepsilon) = \sum_\alpha e_{\alpha i} e_{\alpha j} f_\alpha^{(eq)}(\mathbf{x} + \beta \mathbf{e}_\alpha \varepsilon, t + \beta\varepsilon)$$
$$+ \varepsilon \sum_\alpha e_{\alpha i} e_{\alpha j} f_\alpha^{(1)}(\mathbf{x} + \beta \mathbf{e}_\alpha \varepsilon, t + \beta\varepsilon) + O(\varepsilon^2). \tag{4.25}$$

By use of Eq. (3.41) to evaluate the terms in Eqs. (4.24) and (4.25), we can write Eq. (4.23) as

$$\left[\frac{\partial}{\partial t}(hu_i) + \frac{\partial}{\partial x_j}\left(hu_i u_j + \frac{1}{2} g h^2 \delta_{ij}\right)\right]_{(\mathbf{x}+\beta\mathbf{e}_\alpha\varepsilon, t+\beta\varepsilon)}$$
$$= F_i\left(\mathbf{x} + \frac{1}{2}\mathbf{e}_\alpha \varepsilon, t + \frac{1}{2}\varepsilon\right) - \varepsilon \sum_\alpha e_{\alpha i} e_{\alpha j} f_\alpha^{(1)} + O(\varepsilon^2). \tag{4.26}$$

Combination of Eqs. (4.17) and (4.18) gives

$$\sum_\alpha e_{\alpha i} e_{\alpha j} f_\alpha^{(1)} = -\frac{e^2 \tau}{3}\left[\frac{\partial (hu_i)}{\partial x_j} + \frac{\partial (hu_j)}{\partial x_i}\right]. \tag{4.27}$$

Finally, notice that for the stationary case there is an initial condition of $u_i = 0$, the above equation becomes

$$\sum_\alpha e_{\alpha i} e_{\alpha j} f_\alpha^{(1)} = 0, \tag{4.28}$$

leading to Eq. (4.26) in the following second-order accurate form,

$$h(\mathbf{x} + \beta \mathbf{e}_\alpha \varepsilon, t + \beta\varepsilon)\frac{\partial}{\partial x_i} h(\mathbf{x} + \beta \mathbf{e}_\alpha \varepsilon, t + \beta\varepsilon) = F_i\left(\mathbf{x} + \frac{1}{2}\mathbf{e}_\alpha \varepsilon, t + \frac{1}{2}\varepsilon\right). \tag{4.29}$$

Substitution of Eq. (4.19) into above equation results in

$$h(\mathbf{x} + \beta \mathbf{e}_\alpha \varepsilon, t + \beta\varepsilon)\frac{\partial}{\partial x_i} h(\mathbf{x} + \beta \mathbf{e}_\alpha \varepsilon, t + \beta\varepsilon) =$$
$$-h(\mathbf{x} + \frac{1}{2}\mathbf{e}_\alpha \varepsilon, t + \frac{1}{2}\varepsilon)\frac{\partial}{\partial x_i} z_b(\mathbf{x} + \frac{1}{2}\mathbf{e}_\alpha \varepsilon, t + \frac{1}{2}\varepsilon). \tag{4.30}$$

When $\beta = 1/2$, after rearranging Eq. (4.30), we obtain

$$\left[h\frac{\partial}{\partial x_i}(h + z_b)\right]_{(\mathbf{x}+\frac{1}{2}\mathbf{e}_\alpha\varepsilon, t+\frac{1}{2}\varepsilon)} = 0. \tag{4.31}$$

Because for the stationary case there is initially

$$h + z_b = \text{constant}, \tag{4.32}$$

in the whole considered domain as shown in Fig. 4.1, Eq. (4.31) holds true. It follows that the centred scheme satisfies the $\mathcal{N}$-property.

When $\beta \neq 1/2$, we take a Taylor expansion to the term on the left-hand side of the equation (4.30) in time and space around point $(\mathbf{x}+\frac{1}{2}\mathbf{e}_\alpha\varepsilon, t+\frac{1}{2}\varepsilon)$,

$$h(\mathbf{x}+\beta\mathbf{e}_\alpha\varepsilon, t+\beta\varepsilon)\frac{\partial}{\partial x_i}h(\mathbf{x}+\beta\mathbf{e}_\alpha\varepsilon, t+\beta\varepsilon) =$$
$$h(\mathbf{x}+\frac{1}{2}\mathbf{e}_\alpha\varepsilon, t+\frac{1}{2}\varepsilon)\frac{\partial}{\partial x_i}h(\mathbf{x}+\frac{1}{2}\mathbf{e}_\alpha\varepsilon, t+\frac{1}{2}\varepsilon) + O([\beta-\frac{1}{2}]\varepsilon). \quad (4.33)$$

From Eq. (4.21), we obtain

$$|\beta-\frac{1}{2}| < \frac{1}{2}. \quad (4.34)$$

This is obviously a small quantity and may be reasonably assumed as the same order as ε,

$$O([\beta-\frac{1}{2}]\varepsilon) \sim O(\varepsilon^2), \quad (4.35)$$

because ε is the time step. Consequently, Eq. (4.33) can be replaced with the following second-order accurate expression,

$$h(\mathbf{x}+\beta\mathbf{e}_\alpha\varepsilon, t+\beta\varepsilon)\frac{\partial}{\partial x_i}h(\mathbf{x}+\beta\mathbf{e}_\alpha\varepsilon, t+\beta\varepsilon) =$$
$$h(\mathbf{x}+\frac{1}{2}\mathbf{e}_\alpha\varepsilon, t+\frac{1}{2}\varepsilon)\frac{\partial}{\partial x_i}h(\mathbf{x}+\frac{1}{2}\mathbf{e}_\alpha\varepsilon, t+\frac{1}{2}\varepsilon). \quad (4.36)$$

Substitution of the above equation into Eq. (4.30) also leads to Eq. (4.31), indicating again that the centred scheme satisfies the $\mathcal{N}$-property at second-order accuracy, which is consistent with the accuracy of the lattice Boltzmann equation. The centred scheme satisfying the $\mathcal{N}$-property is further confirmed by the numerical tests, demonstrating the importance of a scheme satisfying the $\mathcal{N}$-property. □

It must be pointed out that we can prove in an exact same manner that neither the basic scheme nor the second-order scheme generally satisfies the $\mathcal{N}$-property, hence producing solutions with large errors to some shallow water flow problems. These are also confirmed by the numerical tests given in Section 7.3.

4.5 Discussions

4.5.1 Properties of Force Term

The force term may depend on both space and time. It can be a constant, linear or non-linear function. For constant force, the three scheme are identical; for linear force term, the centred and the second-order schemes are equivalent; but for non-linear force, only the centred scheme can produce an accurate solution. In general, if the force term is a continuous and first-order differentiable function, both the basic and the second-order schemes may generate good approximations to accurate solutions which can be obtained with the centred scheme. However, if the force function is only continuous but not differentiable, only the centred scheme provides an accurate solution, and the other two provide solutions with various levels of error. Often, the second-order scheme generates worse predictions. This indicates that *a right mathematical formulation may not guarantee its proper approach to a real life problem in physics. Only if the formulation also represents the underlying physics, can it result in a correct solution.*

4.5.2 Recommendation

In Section 4.3, we use the implicit form of the centred scheme (4.4) to show that the scheme is second-order accurate in space and time. Furthermore, we demonstrate in the above section that the scheme satisfies the $\mathcal{N}$-property. Similarly, we can prove that the semi-implicit form of the centred scheme (4.5) is second-order accurate in space but only first-order accurate in time. Since the time step in the lattice Boltzmann method is usually small, a scheme with first-order accuracy in time can still generate an accurate solution to most flow problems. In fact, this has been confirmed in the numerical computations. Hence, Eq. (4.5) is recommended in practical computations for efficiency.

4.5.3 Discretization Errors

Since the centred scheme can produce an accurate solution, we may take it as a staring point to analysis the discretization errors for both the basic and the second-order schemes here.

For simplicity but without loss of generality, we use the schemes in explicit forms in time, i.e. at time level t, in the analysis. By taking a Taylor expansion to the basic scheme (4.1) around point $(\mathbf{x} + \frac{1}{2}\mathbf{e}_\alpha \Delta t, t)$, we obtain

$$F_i = F_i(\mathbf{x}, t) = F_i(\mathbf{x} + \frac{1}{2}\mathbf{e}_\alpha \Delta t, t) - \frac{\Delta t}{2} e_{\alpha j} \frac{\partial F_i(\xi_0, t)}{\partial x_j}, \tag{4.37}$$

where $\mathbf{x} < \xi_0 < (\mathbf{x} + \frac{1}{2}\mathbf{e}_\alpha \Delta t)$. Apparently, $e_{\alpha j}$ has the same order as e, i.e.

$$e_{\alpha j} \sim e. \tag{4.38}$$

From the definition, we have

$$e\Delta t = \Delta x, \tag{4.39}$$

and hence Eq. (4.37) can also be written as

$$F_i = F_i(\mathbf{x}, t) = F_i(\mathbf{x} + \frac{1}{2}\mathbf{e}_\alpha \Delta t, t) + O(\Delta x). \tag{4.40}$$

Similarly, taking a Taylor expansion to the second-order scheme (4.3) around point $(\mathbf{x}+\frac{1}{2}\mathbf{e}_\alpha\Delta t, t)$ leads to

$$F_i = \frac{1}{2}[F_i(\mathbf{x},t) + F_i(\mathbf{x}+\mathbf{e}_\alpha\Delta t, t)] = F_i(\mathbf{x}+\frac{1}{2}\mathbf{e}_\alpha\Delta t, t) + \frac{\Delta t}{4}e_{\alpha j}\left[\frac{\partial F_i(\xi_2,t)}{\partial x_j} - \frac{\partial F_i(\xi_1,t)}{\partial x_j}\right]. \quad (4.41)$$

where $\mathbf{x} < \xi_1 < (\mathbf{x}+\frac{1}{2}\mathbf{e}_\alpha\Delta t)$ and $(\mathbf{x}+\frac{1}{2}\mathbf{e}_\alpha\Delta t) < \xi_2 < (\mathbf{x}+\mathbf{e}_\alpha\Delta t)$. The above equation can also be written as

$$F_i = \frac{1}{2}[F_i(\mathbf{x},t) + F_i(\mathbf{x}+\mathbf{e}_\alpha\Delta t, t)] = F_i(\mathbf{x}+\frac{1}{2}\mathbf{e}_\alpha\Delta t, t) + O(\Delta x), \quad (4.42)$$

if

$$\frac{\partial F_i(\xi_1,t)}{\partial x_j} \neq \frac{\partial F_i(\xi_2,t)}{\partial x_j}. \quad (4.43)$$

Now we can analysis the discretization errors in detail as follows:

- If F_i is a constant, we have
$$\frac{\partial F_i}{\partial x_j} = 0. \quad (4.44)$$
The use of the above expression in Eqs. (4.37) and (4.41) results in that both the basic scheme and the second-order scheme are identical to the centred scheme, indicating that there is no discretization errors in this case.
- If F_i is a linear function, there is
$$\frac{\partial F_i}{\partial x_j} = \text{constant} = C_0. \quad (4.45)$$
Substitution of the above equation into Eq. (4.37) yields
$$F_i = F_i(\mathbf{x},t) = F_i(\mathbf{x}+\frac{1}{2}\mathbf{e}_\alpha\Delta t, t) - \frac{\Delta t}{2}e_{\alpha j}C_0. \quad (4.46)$$
With Eq. (4.38) and (4.39), we can write Eq. (4.46) as
$$F_i = F_i(\mathbf{x},t) = F_i(\mathbf{x}+\frac{1}{2}\mathbf{e}_\alpha\Delta t, t) - \frac{C_0\Delta x}{2}, \quad (4.47)$$
which indicates that there is discretization error for the basic scheme in the order of $O\left(C_0\Delta x/2\right)$. For the second-order scheme, putting Eq. (4.45) into Eq. (4.41) gives
$$F_i = \frac{1}{2}[F_i(\mathbf{x},t) + F_i(\mathbf{x}+\mathbf{e}_\alpha\Delta t, t)] = F_i(\mathbf{x}+\frac{1}{2}\mathbf{e}_\alpha\Delta t, t). \quad (4.48)$$
The right hand side of the above equations is just the centred scheme, meaning that there is no discretization error in the situation.

- If F_i is a non-linear function, Eqs. (4.40) and (4.42) suggest that both the basic scheme and the second-order scheme have the discretization error in typical order of $O(\Delta x)$ with respect to the centred scheme. The real order of the discretization error in this case for the two schemes will also depend on the detail of the force term F_i, for instance, F_i is a non-linear function including a gradient that will introduce an additional discretization error.

It should be noted that the above analysis further confirms the general statements made in Section 4.5.1.

4.6 Closure

The centred scheme is described for the accurate evaluation of the force term in the lattice Boltzmann equation. This greatly enhances the capability of the lattice Boltzmann method and opens many new applications of the methods in simulating complex fluid flows. The use of the semi-implicit form of the centred scheme (4.5) retains the parallel feature of the lattice Boltzmann method. The $\mathcal{N}$-property is introduced which can be used as a basic criterion to justify a numerical scheme for its correctness, hence avoiding a scheme which seems to be mathematically correct but physically improper. This property is in fact a fundamental rule for formulating a new numerical scheme.

5

Turbulence Modelling

5.1 Introduction

Most flows in nature are turbulent. How to model turbulent flows plays an important role in practical engineering. Theoretically, turbulent flows can be described by either the space-filtered flow equations or the time-averaged flow equations. Numerical results show that the space-filtered flow equations are more accurate for turbulent flows and preferred [20]. Based on these equations, the governing equations for turbulent shallow water flows are firstly derived. Then, a lattice Boltzmann model for the shallow water equations with turbulence modelling (LABSWE^{TM}) is described. The main feature of the model is that the flow turbulence is efficiently and naturally taken into account by incorporating the standard subgrid-scale stress model into the lattice Boltzmann equation in a consistent manner with the lattice gas dynamics.

5.2 SGS for Shallow Water Equations

As indicated in Section 2.3, the subgrid-scale stress model is used for modelling flow turbulence. In order to consider the turbulence in shallow water flows, the effect of the flow turbulence must be taken into account in the flow equations. By using the similar procedure to that given in Section 2.4, based on the equations (2.4) and (2.9), we can derive the shallow water equations with the effect of the flow turbulence as

$$\frac{\partial h}{\partial t} + \frac{\partial (hu_j)}{\partial x_j} = 0, \tag{5.1}$$

$$\frac{\partial (hu_i)}{\partial t} + \frac{\partial (hu_iu_j)}{\partial x_j} = -g\frac{\partial}{\partial x_i}(\frac{h^2}{2}) + (\nu + \nu_e)\frac{\partial^2 (hu_i)}{\partial x_j \partial x_j} + F_i. \tag{5.2}$$

It should be pointed out that (1) although we use the same symbols in the above equations as that in Eqs. (2.55) and (2.56) for clarity, the difference exists, i.e. u_i in the above equations is the depth-averaged space-filtered velocity component, (2) the depth-averaged subgrid-scale stress τ_{ij} with eddy viscosity is defined by

$$\tau_{ij} = -\nu_e \left[\frac{\partial (h\tilde{u}_i)}{\partial x_j} + \frac{\partial (h\tilde{u}_j)}{\partial x_i} \right], \tag{5.3}$$

and (3) the eddy viscosity ν_e retains the same form as Eq. (2.10), i.e.

$$\nu_e = (C_s l_s)^2 \sqrt{S_{ij} S_{ij}}, \tag{5.4}$$

but the S_{ij} is replaced with

$$S_{ij} = \frac{1}{2h} \left[\frac{\partial (hu_i)}{\partial x_j} + \frac{\partial (hu_j)}{\partial x_i} \right]. \tag{5.5}$$

5.3 LABSWE^{TM}

Now, we describe a lattice Boltzmann model for the shallow water equations with turbulence modelling (LABSWE^{TM}) by extending the LABSWE described in Chapter 3. Mathematically, if we compare the turbulent shallow water equations (5.1) and (5.2) with the equations (2.55) and (2.56) without flow turbulence, we will notice that the only difference between them lies in the momentum equations, or more specifically, the eddy viscosity ν_e appearing in Eq. (5.2) does not exist in Eq. (2.56). Since the kinematic viscosity ν is determined only by the relaxation time via Eq. (3.66), this suggests that if we redefine a new relaxation time τ_t as

$$\tau_t = \tau + \tau_e, \tag{5.6}$$

which gives a total viscosity ν_t,

$$\nu_t = \nu + \nu_e, \tag{5.7}$$

the solution of the following lattice Boltzmann equation,

$$f_\alpha(\mathbf{x} + \mathbf{e}_\alpha \Delta t, t + \Delta t) - f_\alpha(\mathbf{x}, t) = -\frac{1}{\tau_t}(f_\alpha - f_\alpha^{eq}) + \frac{\Delta t}{6e^2} e_{\alpha i} F_i, \tag{5.8}$$

can generate the solution to the shallow water equations (5.1) and (5.2). This is the basic idea behind the lattice Boltzmann subgrid-scale stress model proposed by Hou et al. [50]. In this way, the flow turbulence is simply and naturally modelled in the standard lattice Boltzmann equation with the total relaxation time τ_t which includes the eddy relaxation time τ_e via Eq. (5.6).

In order to decide the total relaxation time τ_t, first of all, we need to determine the strain-rate tensor S_{ij}. Since the lattice Boltzmann method is characterized by simplicity and efficiency, S_{ij} defined by Eq. (5.5) involves calculation of derivatives and hence is not a suitable form in use. To keep the consistent feature with the lattice gas dynamics, it is natural to calculate S_{ij} in terms of the distribution function. By using the Chapman-Enskog expansion, it can be found that the strain-rate tensor S_{ij} is related to the non-equilibrium momentum flux tensor (see Section 5.4 for detail). This provides a simple and efficient way to calculate S_{ij},

$$S_{ij} = -\frac{3}{2e^2 h\tau_t \Delta t} \sum_\alpha e_{\alpha i} e_{\alpha j} (f_\alpha - f_\alpha^{eq}). \tag{5.9}$$

Then, if assuming ν_t and τ_t also satisfy the relation (3.66), we have

$$\tau_t = \frac{1}{2} + \frac{3\nu_t}{e^2 \Delta t}. \tag{5.10}$$

Substitution of Eqs. (5.6) and (5.7) into above equation gives

$$\tau_e + \tau = \frac{1}{2} + \frac{3(\nu_e + \nu)}{e^2 \Delta t}. \tag{5.11}$$

With reference to Eq. (3.66), this is simplified as

$$\tau_e = \frac{3}{e^2 \Delta t} \nu_e. \tag{5.12}$$

Now, replacing ν_e in the above equation with Eq. (5.4) leads to

$$\tau_e = \frac{3}{e^2 \Delta t} (C_s l_s)^2 \sqrt{S_{ij} S_{ij}}. \tag{5.13}$$

Substituting Eq. (5.9) into above equation we obtain

$$\tau_e = \frac{3}{e^2 \Delta t} (C_s l_s)^2 \frac{3}{2e^2 h \tau_t \Delta t} \sqrt{\Pi_{ij} \Pi_{ij}}, \tag{5.14}$$

where

$$\Pi_{ij} = \sum_\alpha e_{\alpha i} e_{\alpha j} (f_\alpha - f_\alpha^{eq}). \tag{5.15}$$

Using relation (5.6), if $l_s = \Delta x$ is adopted, we can further write the equation (5.14) as

$$\tau_e = \frac{9}{2} \frac{C_s^2}{e^2 h (\tau_e + \tau)} \sqrt{\Pi_{ij} \Pi_{ij}}. \tag{5.16}$$

Finally, solution of the above equation results in the eddy relaxation time,

$$\tau_e = \frac{-\tau + \sqrt{\tau^2 + 18 C_s^2 / (e^2 h) \sqrt{\Pi_{ij} \Pi_{ij}}}}{2}, \tag{5.17}$$

hence giving the following total relaxation time τ_t via Eq. (5.6),

$$\tau_t = \frac{\tau + \sqrt{\tau^2 + 18 C_s^2 / (e^2 h) \sqrt{\Pi_{ij} \Pi_{ij}}}}{2}. \tag{5.18}$$

5.4 Recovery of Turbulent SWE

In the similar manner as that described in Section 3.6, we can show that the lattice Boltzmann equation (5.8) results in solution to the macroscopic equations (5.1) and (5.2) by use of the Chapman-Enskog expansion. After assuming Δt is small and is equal to ε,

$$\Delta t = \varepsilon, \tag{5.19}$$

we write the equation (5.8) as

$$f_\alpha(\mathbf{x} + \mathbf{e}_\alpha \varepsilon, t + \varepsilon) - f_\alpha(\mathbf{x}, t) = -\frac{1}{\tau_t}(f_\alpha - f_\alpha^{eq}) + \frac{\varepsilon}{6e^2} e_{\alpha j} F_j. \tag{5.20}$$

Taking a Taylor expansion to the first term on the left-hand side of the above equation in time and space around point $(\mathbf{x}, t)$ leads to

$$\varepsilon(\frac{\partial}{\partial t} + e_{\alpha j}\frac{\partial}{\partial x_j}) f_\alpha + \frac{1}{2}\varepsilon^2(\frac{\partial}{\partial t} + e_{\alpha j}\frac{\partial}{\partial x_j})^2 f_\alpha + \mathcal{O}(\varepsilon^2) = -\frac{1}{\tau_t}(f_\alpha - f_\alpha^{(0)}) + \frac{\varepsilon}{6e^2} e_{\alpha j} F_j. \tag{5.21}$$

Also, we expand f_α around $f_\alpha^{(0)}$, i.e.

$$f_\alpha = f_\alpha^{(0)} + \varepsilon f_\alpha^{(1)} + \varepsilon^2 f_\alpha^{(2)} + \mathcal{O}(\varepsilon^2), \tag{5.22}$$

where $f_\alpha^{(0)} = f_\alpha^{eq}$.

The equation (5.21) to order ε is

$$(\frac{\partial}{\partial t} + e_{\alpha j}\frac{\partial}{\partial x_j}) f_\alpha^{(0)} = -\frac{1}{\tau_t} f_\alpha^{(1)} + \frac{1}{6e^2} e_{\alpha j} F_j \tag{5.23}$$

and to order ε^2 is

$$(\frac{\partial}{\partial t} + e_{\alpha j}\frac{\partial}{\partial x_j}) f_\alpha^{(1)} + \frac{1}{2}(\frac{\partial}{\partial t} + e_{\alpha j}\frac{\partial}{\partial x_j})^2 f_\alpha^{(0)} = -\frac{1}{\tau_t} f_\alpha^{(2)}. \tag{5.24}$$

Substitution of Eq. (5.23) into Eq. (5.24) gives

$$(1 - \frac{1}{2\tau_t})(\frac{\partial}{\partial t} + e_{\alpha j}\frac{\partial}{\partial x_j}) f_\alpha^{(1)} = -\frac{1}{\tau_t} f_\alpha^{(2)} - \frac{1}{2}(\frac{\partial}{\partial t} + e_{\alpha j}\frac{\partial}{\partial x_j})(\frac{1}{6e^2} e_{\alpha k} F_k). \tag{5.25}$$

Taking $\sum$ [(5.23) + $\varepsilon\times$ (5.25)] about α provides

$$\frac{\partial}{\partial t}(\sum_\alpha f_\alpha^{(0)}) + \frac{\partial}{\partial x_j}(\sum_\alpha e_{\alpha j} f_\alpha^{(0)}) = -\varepsilon \frac{1}{12e^2}\frac{\partial}{\partial x_j}(\sum_\alpha e_{\alpha j} e_{\alpha k} F_k). \tag{5.26}$$

Again, if the first-order accuracy for the force term is applied, evaluation of the other terms in the above equation using Eq. (3.41) results in the continuity equation (5.1).

From $\sum e_{\alpha i}$[(5.23) + $\varepsilon\times$ (5.25)] about α, we have

$$\frac{\partial}{\partial t}(\sum_\alpha e_{\alpha i} f_\alpha^{(0)}) + \frac{\partial}{\partial x_j}(\sum_\alpha e_{\alpha i} e_{\alpha j} f_\alpha^{(0)}) + \varepsilon(1 - \frac{1}{2\tau_t})\frac{\partial}{\partial x_j}(\sum_\alpha e_{\alpha i} e_{\alpha j} f_\alpha^{(1)})$$
$$= F_j \delta_{ij} - \varepsilon \frac{1}{2} \sum_\alpha e_{\alpha i} (\frac{\partial}{\partial t} + e_{\alpha j}\frac{\partial}{\partial x_j})(\frac{1}{6e^2} e_{\alpha j} F_j). \tag{5.27}$$

Also, if the first-order accuracy for the force term is used, after the other terms is simplified with Eq. (3.41), the above equation becomes

$$\frac{\partial (hu_i)}{\partial t} + \frac{\partial (hu_i u_j)}{\partial x_j} = -g\frac{\partial}{\partial x_i}(\frac{h^2}{2}) - \frac{\partial}{\partial x_j}\Lambda_{ij} + F_i, \tag{5.28}$$

where

$$\Lambda_{ij} = \frac{\varepsilon}{2\tau_t}(2\tau_t - 1)\sum_\alpha e_{\alpha i} e_{\alpha j} f_\alpha^{(1)}. \tag{5.29}$$

With reference to Eq. (5.23), using Eq. (3.41), after some algebra, we obtain

$$\Lambda_{ij} \approx -\frac{\varepsilon}{6} e^2 (2\tau_t - 1)\left[\frac{\partial (hu_i)}{\partial x_j} + \frac{\partial (hu_j)}{\partial x_i}\right]. \tag{5.30}$$

Substitution of Eq. (5.30) into Eq. (5.28) results in a momentum equation,

$$\frac{\partial (hu_i)}{\partial t} + \frac{\partial (hu_i u_j)}{\partial x_j} = -g\frac{\partial}{\partial x_i}(\frac{h^2}{2}) + \nu_t \frac{\partial^2 (hu_i)}{\partial x_j \partial x_j} + F_i, \tag{5.31}$$

with the total viscosity defined by

$$\nu_t = \frac{e^2 \Delta t}{6}(2\tau_t - 1). \tag{5.32}$$

Notice Eqs. (3.66), (5.6), (5.7) and (5.12), the above total viscosity becomes

$$\nu_t = \nu_e + \nu, \tag{5.33}$$

and hence Eq. (5.31) is just the momentum equation (5.2)

It should be pointed out that the use of the centred scheme for the force term described in Section 4 can generate the second-order accurate macroscopic continuity and momentum equations in time and space.

Now, we derive the relation (5.9) for the strain-rate tensor S_{ij}. By combining of Eqs. (5.29) and (5.30), we have

$$\frac{1}{2h}\left[\frac{\partial (hu_i)}{\partial x_j} + \frac{\partial (hu_j)}{\partial x_i}\right] = -\frac{3}{2e^2 h \tau_t}\sum_\alpha e_{\alpha i} e_{\alpha j} f_\alpha^{(1)}. \tag{5.34}$$

From Eq. (5.22), there is

$$f_\alpha^{(1)} = \frac{(f_\alpha - f_\alpha^{(0)})}{\varepsilon} + \mathcal{O}(\varepsilon) \approx \frac{(f_\alpha - f_\alpha^{(0)})}{\varepsilon}. \tag{5.35}$$

Notice $f_\alpha^{(0)} = f_\alpha^{eq}$, $\varepsilon = \Delta t$ and the definition for strain-rate tensor S_{ij} by Eq. (5.5), substitution of Eq. (5.35) into Eq. (5.34) leads to Eq. (5.9).

5.5 Closure

This chapter describes the lattice Boltzmann model for the shallow water equations with turbulence modelling (LABSWE^{TM}). The idea behind the model is to introduce the eddy relaxation time τ_e to the collision term. The standard Smagorinsky SGS model is included in a natural and efficient way for flow turbulence. This makes it extremely simple to model turbulent flows by means of the lattice Boltzmann method. The model has potential capability and may be used to solve practical flow problems.

6

Boundary and Initial Conditions

6.1 Introduction

The lattice Boltzmann models for the shallow water equations with or without turbulence modelling (LABSWE and LABSWE^{TM}) are described in Chapters 3 and 5, respectively. The centred scheme for accurate evaluation of the force term in the lattice Boltzmann equation is described in Chapter 4. In order to solve shallow water flow problems by use of either LABSWE or LABSWE^{TM}, suitable boundary conditions must be provided. Therefore, this chapter firstly describes various boundary conditions for solid walls, which cover the bounce-back scheme for no-slip boundary conditions and the elastic-collision scheme for slip and semi-slip boundary conditions. Then, the boundary conditions for inflow and outflow boundaries are introduced. Finally, initial conditions are discussed.

6.2 Bounce-Back Scheme

The lattice Boltzmann method is a very promising computational method. From the point of view of the capability of treating boundary conditions, the lattice Boltzmann method has two major advantages over the other methods, i.e. (1) the uniform lattice which is generally suitable for arbitrary complex geometries without any conventional mesh generation, and (2) the bounce-back scheme which is most efficient for implementation of boundary conditions. It is these advantages that make the lattice Boltzmann method most efficient in simulating flows with arbitrary complex geometries such as flows through porous media.

The basic idea behind the bounce-back scheme is very simple and stats that an incoming particle towards the boundary is bounce back into fluid. Now let's look at the boundary shown in Fig. 6.1. The incoming known distribution functions f_6, f_7 and f_8 move towards the solid wall and they are immediately bounced back by the wall, i.e. the unknown f_2, f_3 and f_4 after streaming can be simply decided as

$$f_2 = f_6, \qquad f_3 = f_7, \qquad f_4 = f_8. \tag{6.1}$$

As a result, the sum of the particle momentum close to solid wall is zero, implying that this bounce-back scheme leads to no-slip boundary conditions. Because of its simplicity and easy implementation, the scheme is the most popular way for no-slip boundary conditions used in the lattice Boltzmann method for simulating fluid flows.

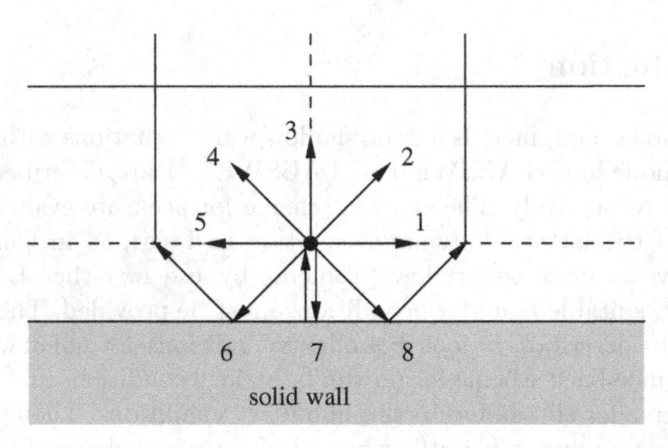

Fig. 6.1. Sketch for bounce-back and elastic-collision schemes.

6.3 Elastic-Collision Scheme

In addition to the above bounce-back scheme, there is another standard scheme, elastic-collision scheme, for boundary conditions in the lattice gas dynamics. This is, however, seldom applied in the lattice Boltzmann method due to its difficulty in dealing with flows within arbitrary complex geometries, which is briefly reviewed as follows.

The standard elastic-collision scheme for boundary conditions states that an incoming particle towards the boundary is reflected back into fluid as shown in Fig. 6.1, i.e. the unknown f_2, f_3 and f_4 after streaming can be decided as

$$f_2 = f_8, \qquad f_3 = f_7, \qquad f_4 = f_6. \tag{6.2}$$

Consequently, the sum of the particle momentum normal to the solid wall is zero, indicating that this elastic-collision scheme leads to slip boundary condition for plane solid wall, which is as efficient and robust as the bounce-back scheme. However, for flows in arbitrary complex geometries, it is impossible to implement this idea to achieve slip boundary conditions. Therefore, in this section, we introduce a general elastic-collision scheme to achieve slip and semi-slip boundary conditions for flows in arbitrary complex geometries. The scheme was developed by Zhou [17].

6.3.1 Representation of Boundary

First of all, a lattice cell is defined as the control volume which is constructed around each lattice node by placing their boundaries midway between the lattice nodes as shown with solid lines in Fig. 6.2. Then, a boundary shape is represented with a series of straight-line segments in an anti-clockwise direction. Finally, we find all the intersect points between the lattice cells and the solid boundaries with the method described by Quirk [51]. As a result, the lattice nodes are divided into flow and solid nodes. In order to treat boundaries, we further define a boundary node which is the flow node with at least one neighboring solid node, and the corresponding lattice cell is called a boundary cell. Depending on the slope θ of the interface between boundaries and lattice cells (see Fig. 6.2), four types of the boundary nodes are defined: (1) $0^0 < \theta \le 90^0$; (2) $90^0 < \theta \le 180^0$; (3) $180^0 < \theta \le 270^0$; and (4) $270^0 < \theta \le 360^0$. Next, we use a variable $C_{i,j}$ to denote each lattice node as flow node with $C_{i,j} = 0$, boundary node with $C_{i,j} = 0.5$ and solid node with $C_{i,j} = 1$. Furthermore, within each type, three subtypes are defined as shown in Fig. 6.3, e.g. we have (i) $0^0 < \theta \le \theta_0$; (ii) $\theta_0 < \theta \le (90^0 - \theta_0)$; and (iii) $(90^0 - \theta_0) < \theta \le 90^0$ for Type 1 ($0^0 < \theta \le 90^0$), where θ_0 is called a characteristic angle which is a constant ($0 < \theta_0 < 45^0$). The subtypes for the left three types can be similarly defined.

6.3.2 Slip Boundary Condition

Based on the three subtypes of the boundary nodes, a general elastic-collision scheme is described for the Type 1 as follows for slip boundary conditions. The procedures for the left three types can be formulated in exactly the same manner as that for Type 1.

1. $0^0 < \theta \le \theta_0$: the interface, straight-line segment $\overline{ad}$ shown in Fig. 6.3(i) is treated as a horizontal line; after streaming, the unknown distribution function f_7 and the possibly unknown distribution functions f_6 and f_8 can be determined as

$$\begin{aligned} f_7 &= f_3, & & \\ f_6 &= f_4, & C_{i+1,j+1} &= 1, \\ f_8 &= f_2, & C_{i-1,j+1} &= 1, \end{aligned} \tag{6.3}$$

which is also suitable for the case where the interface is in the neighboring north cell of the boundary cell, e.g. the interface $\overline{ab}$ for the boundary node $(i-1, j-1)$ in Fig. 6.2.

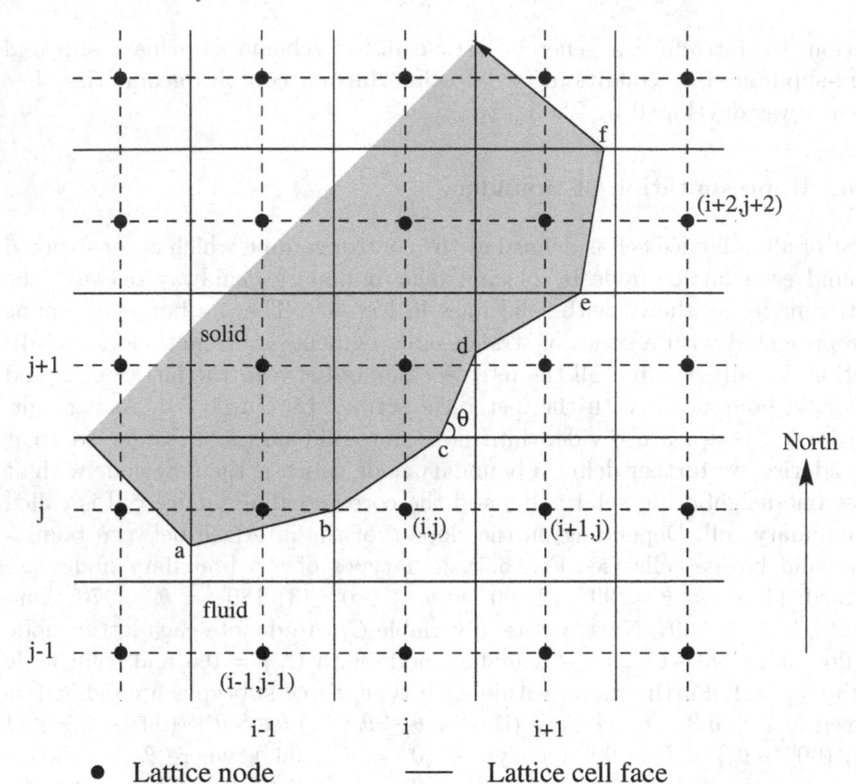

Fig. 6.2. Definition sketch for lattice nodes and lattice cells.

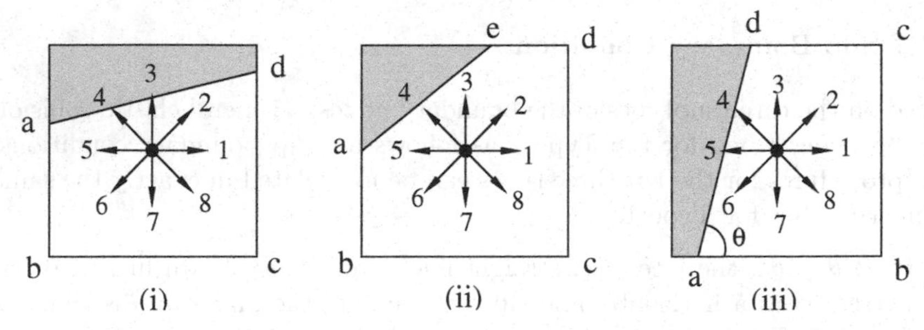

Fig. 6.3. Three subtypes for Type 1: (i) $0^0 < \theta \le \theta_0$; (ii) $\theta_0 < \theta \le (90^0 - \theta_0)$; and (iii) $(90^0 - \theta_0) < \theta \le 90^0$.

2. $\theta_0 < \theta \le (90^0 - \theta_0)$: the interface, straight-line segment $\overline{ae}$ in Fig. 6.3(ii) is regarded as an inclined line of $\theta = 45^0$; again, after streaming, the unknown distribution function f_8 and the possibly unknown distribution functions f_1 and f_7 are specified by

$$f_8 = f_4,$$
$$f_7 = f_5, \qquad C_{i,j+1} = 1, \tag{6.4}$$
$$f_1 = f_3, \qquad C_{i-1,j} = 1.$$

3. $(90^0 - \theta_0) < \theta \leq 90^0$: the interface, straight-line segment $\overline{ad}$ in Fig. 6.3(iii) is approximated with a vertical line; also, the unknown distribution function f_1 and the possibly unknown distribution functions f_2 and f_8 are decided as

$$f_1 = f_5,$$
$$f_8 = f_6, \qquad C_{i-1,j+1} = 1, \tag{6.5}$$
$$f_2 = f_4, \qquad C_{i-1,j-1} = 1,$$

which is also applied to the case where the interface is in the neighboring west cell of the boundary cell, e.g. the interface $\overline{ef}$ for the boundary node $(i+2, j+2)$ in Fig. 6.2.
4. one special case: when a boundary node has only one northwest solid node which is in the same lattice cell as the interface such as the interface $\overline{cd}$ for the boundary node $(i+1, j)$ (see Fig. 6.2), there is only one unknown distribution function f_8 which can be decided with the bounce-back scheme as

$$f_8 = f_4, \tag{6.6}$$

no matter what the slope θ of the interface is.

It must be noted that the characteristic angle θ_0 is not yet set to a value. The effect of θ_0 on boundary conditions was investigated numerically by Zhou [17]. The study shows that slip velocities are effectively independent of θ_0 if $15^0 \leq \theta_0 \leq 30^0$, which is a highly desirable feature of the scheme. The reasons for this are: (a) the macroscopic properties are the aggregate effect of all the microscopic interactions, and the former hydrodynamics depends very little on the details of the latter dynamics, consistent with the theory of lattice gas automata; (b) θ_0 firstly decides a main lattice link which is closest to a normal line to boundary, e.g. Link 3 for Subtype (i) in Fig. 6.3, and then only one unknown distribution function such as f_7 along its opposite link (Link 7) needs to be determined; and (c) the possibly unknown distribution functions such as f_6 and f_8 along the links (Links 6, 8) opposite the neighboring links (Links 2, 4) are decided according to the real situation of their neighboring nodes, which smooth the change from Subtype 1 to Subtype 2, to Subtype 3. For practical computations, $\theta_0 = 20^0$ is recommended.

6.3.3 Semi-Slip Boundary Condition

Physically, there is a large gradient in the vicinity of solid boundary for turbulent flows because of wall friction. This cannot be correctly modeled with no-slip boundary conditions, but is better represented with semi-slip boundary conditions in a mathematical model, which is of great interest in practical simulations. To achieve a semi-slip boundary condition, the effect of wall shear stress must be taken into account. The wall shear stress vector $\boldsymbol{\tau}_f$ due to wall friction may be expressed as [52]

$$\boldsymbol{\tau}_f = -\rho\nu\frac{\partial \mathbf{V}_\tau}{\partial n} = -\rho C_f|\mathbf{V}_\tau|\mathbf{V}_\tau, \tag{6.7}$$

where $\mathbf{V}_\tau$ is velocity vector parallel to the wall; n is the outward coordinate normal to the wall; and C_f is the friction factor at the wall which may be either constant or estimated by

$$C_f = g\frac{n_f^2}{h^{1/3}}, \tag{6.8}$$

in which n_f is the Manning's coefficient at the wall.

If the slip boundary condition described in the previous section is used at the boundary node, the velocity vector normal to the wall tends to be zero, which indicates that for a boundary node we have

$$\mathbf{V}_\tau \approx \mathbf{V}. \tag{6.9}$$

Substitution of Eq. (6.9) into Eq. (6.7) leads to the following expression in a tensor form,

$$\tau_{fi} = -\rho C_f u_i\sqrt{u_j u_j}. \tag{6.10}$$

Therefore, a natural and simple way for semi-slip boundary condition is to solve the lattice Boltzmann equation of either Eq. (3.16) or Eq. (5.8) at the boundary nodes by adding the wall shear stress to the force term F_i, i.e.

$$F_i = -gh\frac{\partial z_b}{\partial x_i} + \frac{\tau_{wi}}{\rho} - \frac{\tau_{bi}}{\rho} + \frac{\tau_{fi}}{\rho} \tag{6.11}$$

together with the slip boundary conditions such as Eqs. (6.3) - (6.6).

6.4 Inflow and Outflow

At the inflow boundary shown in Fig. 6.4, the distribution functions, f_1, f_2 and f_8, at the lattice nodes along Line $\overline{AD}$ cannot be determined with that from the internal lattice nodes. They must be determined with suitable boundary conditions. In practical computations, we found that setting the zero gradient of the distribution function normal to the boundary is often satisfactory, i.e. after streaming, the unknown f_1, f_2 and f_8 are simply calculated by

$$f_\alpha(1, j) = f_\alpha(2, j), \qquad \alpha = 1,\ 2,\ 8. \tag{6.12}$$

Similarly, we can have the following relations for f_4, f_5 and f_6 at the outflow boundary $\overline{BC}$ (see Fig. 6.4),

$$f_\alpha(N_x, j) = f_\alpha(N_x - 1, j), \qquad \alpha = 4,\ 5,\ 6, \tag{6.13}$$

where N_x is the total lattice number in x direction. The correct boundary conditions for the macroscopic variables such as velocities are specified in a normal way. Since the local equilibrium function is the function of the macroscopic variables, a correct distribution function at the boundaries will be accordingly generated in this way.

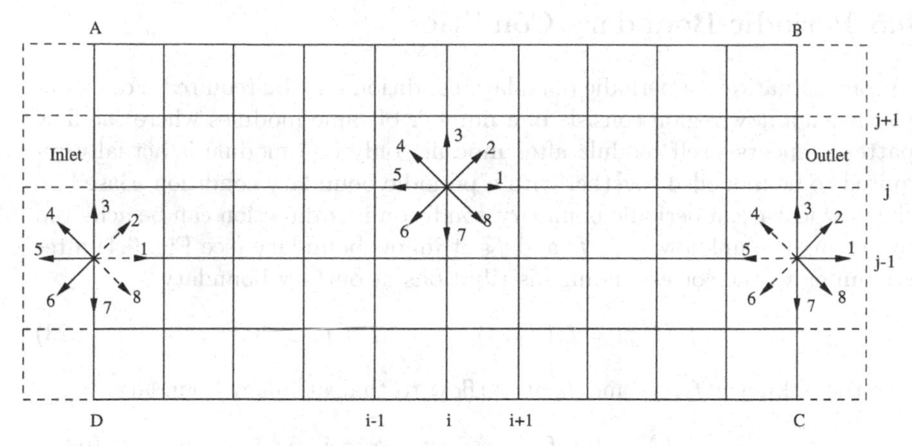

Fig. 6.4. Definition sketch for lattice nodes at inflow and outflow boundaries.

Also, if the velocity and the depth are known, the unknown distribution function f_α at the boundary can be decided with the method described by Zou and He [53]. At the inflow boundary, by using the relations (3.46) and (3.51), we obtian three equations,

$$f_1 + f_2 + f_8 + f_3 + f_4 + f_5 + f_6 + f_7 + f_9 = h, \tag{6.14}$$

$$e(f_1 + f_2 + f_8) - e(f_4 + f_5 + f_6) = hu, \tag{6.15}$$

$$e(f_2 + f_4) - e(f_6 + f_8) + ef_3 + ef_7 = hv. \tag{6.16}$$

If $v = 0$ is assumed, solution of the above three equations for f_1, f_2 and f_8 results in

$$f_1 = f_5 + \frac{2hu}{3e}, \tag{6.17}$$

$$f_2 = \frac{hu}{6e} + f_6 + \frac{f_7 - f_3}{2}, \tag{6.18}$$

$$f_8 = \frac{hu}{6e} + f_4 + \frac{f_3 - f_7}{2}. \tag{6.19}$$

Following the same procedure, we can determine the unknown f_4, f_5 and f_6 for outflow boundary as

$$f_5 = f_1 - \frac{2hu}{3e}, \tag{6.20}$$

$$f_4 = -\frac{hu}{6e} + f_8 + \frac{f_7 - f_3}{2}, \tag{6.21}$$

$$f_6 = -\frac{hu}{6e} + f_2 + \frac{f_3 - f_7}{2}. \tag{6.22}$$

6.5 Periodic Boundary Condition

In some situations, a periodic boundary condition may be required. For example, when a flow region consists of a number of same modules where the flow pattern repeats itself module after module, only one module is actually required to be modelled together with a periodic boundary condition. Based on the flow feature, a periodic boundary condition in x direction can be achieved by setting the unknown f_1, f_2 and f_8 at inflow boundary (see Fig. 6.4) after streaming to the corresponding distributions at outflow boundary,

$$f_\alpha(1, j) = f_\alpha(N_x, j), \qquad \alpha = 1,\ 2,\ 8, \tag{6.23}$$

and the unknown f_4, f_5 and f_6 at outflow to that at inflow boundary,

$$f_\alpha(N_x, j) = f_\alpha(1, j), \qquad \alpha = 4,\ 5,\ 6. \tag{6.24}$$

Similarly, a periodic boundary condition in y direction can be formulated.

6.6 Initial Condition

In general, there are two ways to specify an initial condition in the lattice Boltzmann method. One is to set random value between 0 and 1 for the distribution function f_α. Another is to define flow field first and calculate the local equilibrium distribution function f_α^{eq} which is then used as an initial condition for f_α, i.e. $f_\alpha = f_\alpha^{eq}$. From the point of view of physics, it is often easier to specify a macroscopic quantity than a microscopic. Hence the second method is preferred in practical computations and is used in the LABSWE and LABSWE^{TM}. Numerical tests indicate that this tends to speed up the computations and generate accurate solutions in most situations. Obviously, there is no difference between solutions obtained with these two initial conditions for a steady flow problem.

6.7 Solution Procedure

The solution method for the LABSWE and LABSWE^{TM} is extremely simple. It involve only explicit calculations and consists of the following procedure:

1. given initial water depth and velocity,
2. calculate f_α^{eq} from Eq. (3.41),
3. compute f_α from the lattice Boltzmann equation (3.16) with the relaxation time τ, or from the equation (5.8) for turbulent flows together with the total relaxation time τ_t calculated from Eq. (5.18),
4. update the depth and the velocity according to Eqs. (3.46) and (3.51),
5. return to step 2 and repeat the above procedure until a solution is obtained.

6.8 Discussions

In general, for steady flows, the time step Δt in the lattice Boltzmann equation plays an equivalent role to iteration number in the computation without any physical interpretation. A steady-state solution is achieved when the sum of the change in either the distribution functions or a physical variable such as water depth or velocity is smaller than a pre-defined convergence criterion.

However, for unsteady flows, the time step Δt represents a portion of the real time in simulating a physical fluid. As a result, a numerical simulation at each time step will reflect the current state of the fluid from the beginning to the current time that is the sum of the time over all the past time steps. Thus, the use of a suitable time step is important for accurate solution to such flow problems. To find a correct time step, a few preliminary tests with several time steps are usually carried out for some reasonable time and the results at the same real time are then compared. If the difference between the two successive time steps is less than a pre-defined criterion, the choice of the smaller time step is likely to produce accurate solution, which may be referred to time-step independent solution to an unsteady flow problem.

6.9 Closure

Various boundary conditions together with initial conditions are described in detail for the lattice Boltzmann methods. This completes the lattice Boltzmann model for shallow water flows, implying that the LABSWE and LABSWE^{TM} are two very powerful models for simulating practical shallow water flows with or without flow turbulence. In the following chapter, we will apply the models to solve different benchmark tests in order to demonstrate their potential capabilities.

7

Applications

7.1 Introduction

In this chapter, we shall present several applications of the lattice Boltzmann methods for the shallow water flows (LABSWE and LABSWE^{TM}) described in this book. The methods have been tested and applied to a variety of flow situations: steady and unsteady flows, tidal flows, turbulent flows and wind-induced circulations. Comparisons of the numerical results with either analytical solutions or available experimental data are provided to demonstrate the accuracy, efficiency and capability of the methods.

7.2 Basic Tests

The numerical results illustrated in this section were carried out by Zhou [15] to assess the LABSWE. This includes a steady flow, a tidal flow and a flow around a cylinder. The complete detail can be found in Reference [15].

7.2.1 Steady Flow over a Bump

A 1D steady subcritical flow in a 25 m long channel with a bump (see Fig. 7.1) defined by

$$z_b(x) = \begin{cases} 0.2 - 0.05(x-10)^2, & 8 < x < 12, \\ 0, & \text{otherwise}, \end{cases} \tag{7.1}$$

is a classical test problem which has been used as a benchmark test case for numerical methods at a workshop on dam-break wave simulations [54]. The problem was also used by Vázquez-Cendón [55] to test their scheme with an upwind discretisation for the bed slope source term. According to the hydraulics, when a steady subcritical flow passes over a bump on bed, there is surface drop above the bump. The analytical solution is given by Goutal [54]. This is used as the first test for the presented methods.

In the numerical computations, the discharge per unit width of $q = 4.42\ m^2/s$ was imposed at the inflow and $h = 2\ m$ was specified at the downstream end. $e = 15\ m/s$ and $\tau = 1.5$. The 1D flow can be guaranteed in the 2D code by specifying slip boundary conditions at side walls, which is achieved with the the elastic-collision scheme given in Section 6.3. Eqs. (6.12) and (6.13) are used for the inflow and outflow boundary conditions. The computations indicate that the results based on lattice with $\Delta x \leq 0.1\ m$ provide lattice-independent solutions. Fig. 7.1 shows the profile of the water surface along the channel, in which the surface drop above the bump is clearly formed because the flow is subcritical. Comparison of the result with the analytical solution is plotted in Fig. 7.2, showing an excellent agreement. The quantitative comparison with the analytical solution indicates that the maximum relative error for the water depth is smaller than 0.25%. In order to test the conservative property of the model, the numerical solution of the discharge (mass) is shown in Fig. 7.3. Also the quantitative comparison with the theoretical discharge shows that the maximum relative error is smaller than 0.15%. This suggests that the LABSWE is accurate and conservative.

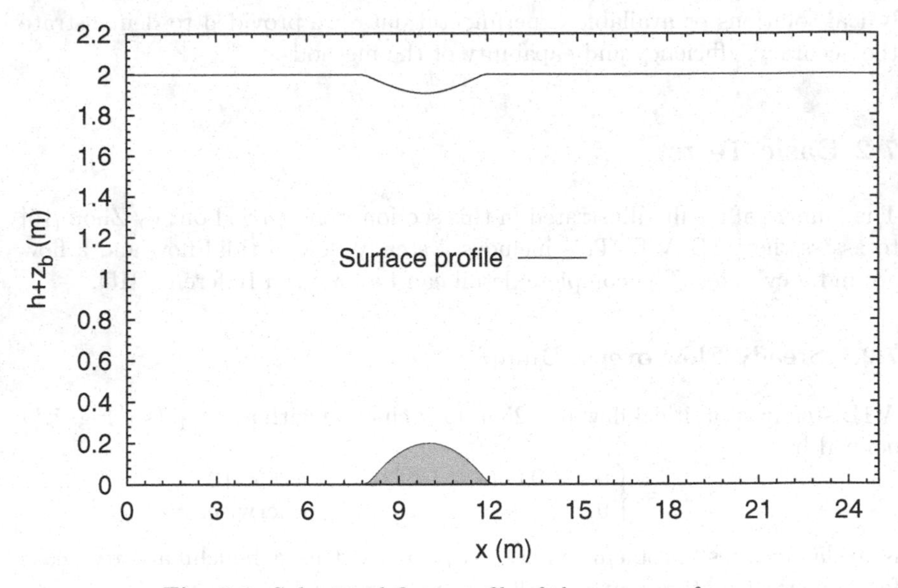

Fig. 7.1. Subcritical flow: profile of the water surface.

7.2.2 Tidal Flow over a Regular Bed

Tidal flows occur in coastal engineering. Here we consider the test problem that Bermudez and Vázquez [56] used for verification of an upwind discretisation of the bed slope source term. This is a one dimensional problem with

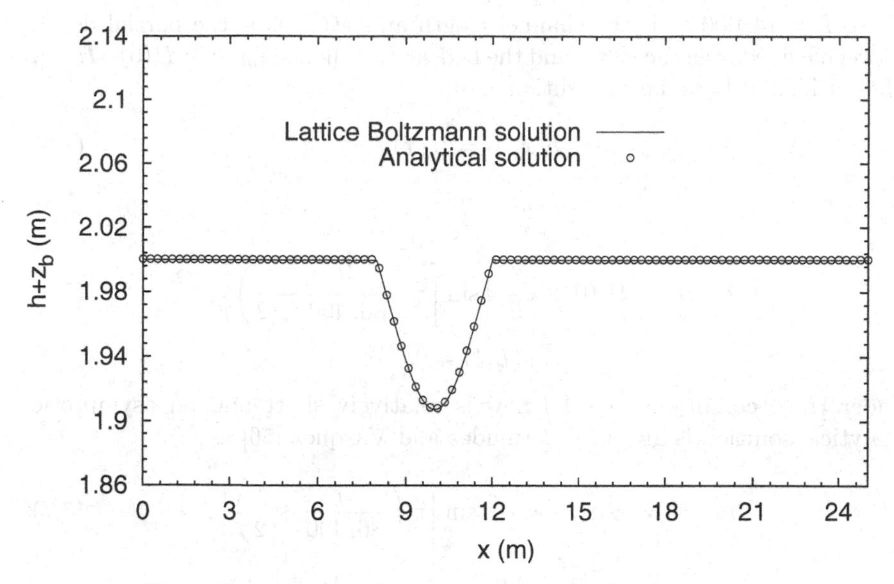

Fig. 7.2. Subcritical flow: comparison of the water surface.

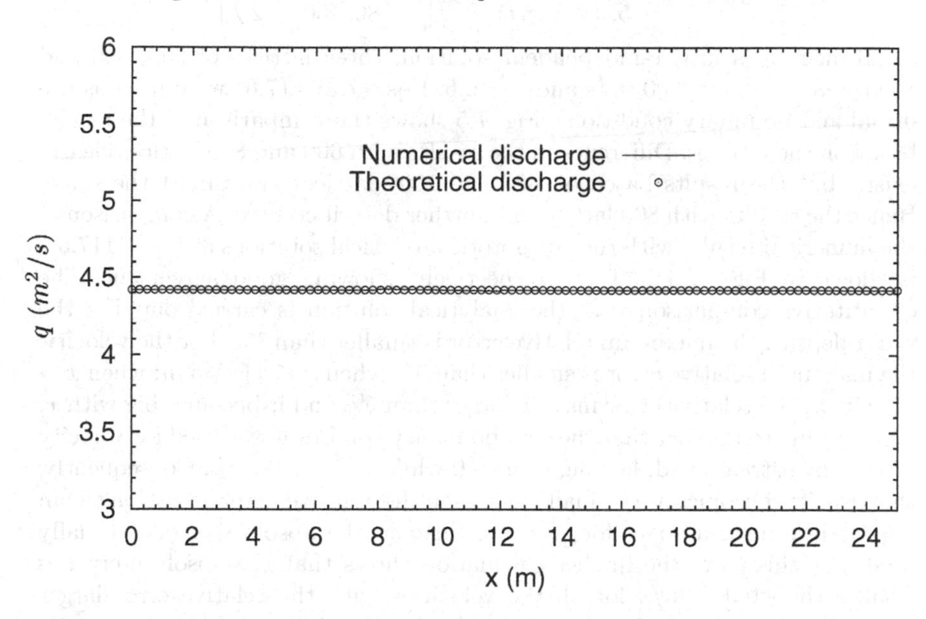

Fig. 7.3. Subcritical flow: comparison of the discharge.

bed topography defined by (see Fig. 7.4)

$$H(x) = 50.5 - \frac{40x}{L} - 10\sin\left[\pi\left(\frac{4x}{L} - \frac{1}{2}\right)\right], \tag{7.2}$$

where $L = 14,000\ m$ is the channel length and $H(x)$ is is the partial depth between a fixed reference level and the bed surface, hence $z_b(x) = H(0) - H(x)$. The initial and boundary conditions are

$$h(x,0) = H(x), \tag{7.3}$$

$$u(x,0) = 0 \tag{7.4}$$

and

$$h(0,t) = H(0) + 4 - 4\sin\left[\pi\left(\frac{4t}{86,400} + \frac{1}{2}\right)\right], \tag{7.5}$$

$$u(L,t) = 0. \tag{7.6}$$

Under these conditions, a tidal flow is relatively short and an asymptotic analytical solution is given by Bermudez and Vázquez [56] as

$$h(x,t) = H(x) + 4 - 4\sin\left[\pi\left(\frac{4t}{86,400} + \frac{1}{2}\right)\right], \tag{7.7}$$

$$u(x,t) = \frac{(x - 14,000)\pi}{5,400h(x,t)}\cos\left[\pi\left(\frac{4t}{86,400} + \frac{1}{2}\right)\right]. \tag{7.8}$$

To achieve a lattice-independent solution, three lattices of 600, 800 and 1000 were used. $e = 200\ m/s$ and $\tau = 0.6$. Eqs. (7.3) - (7.6) were used as the initial and boundary conditions. Fig. 7.5 shows the comparison of the results based on the lattices. Difference in the results with 600 and 800 lattices clearly exists, but the results based on 800 and 1000 lattices are almost the same. Hence the results with 800 lattices are further described here. A comparison of the numerical results with the asymptotic analytical solutions at $t = 9,117.5\ s$ is shown in Figs. 7.4 and 7.5, respectively, showing good agreement. The quantitative comparison with the analytical solution is carried out. For the water depths, the maximum relative error is smaller than 1%. For the velocity, the maximum relative error is smaller than 5% when $x \leq 11,445\ m$; when $x > 11,445\ m$, the relative error may be larger than 5% and it becomes big with x, which is due to the fact that the zero boundary condition was used for velocity at the downstream end, leading to $u \to 0$ while $x \to 14,000\ m$. Consequently, the velocity becomes very small such that the relative error is no longer an appropriate measurement for accuracy; instead the absolute error is usually used. For this case, the further calculation shows that the absolute error is smaller than $0.006\ m/s$ for all the velocities with the relative error larger than 5%. Since the analytical solution is obtained by use of the asymptotic analysis, such agreement may be considered to be excellent. This confirms that the model is accurate for unsteady shallow water flow problems. The present method can provide solution of the same accuracy as that reported by Bermudez and Vázquez with a high-resolution Godunov-type method [56].

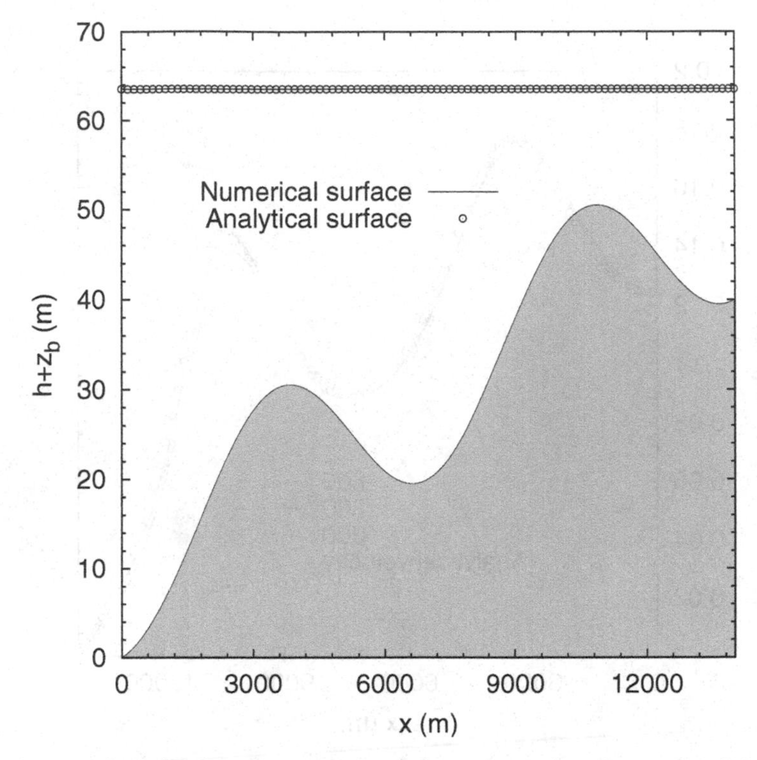

Fig. 7.4. Tidal flow: comparison of the water surface at $t = 9,117.5$ s.

7.2.3 Flow around a Cylinder

Flow around a cylinder is a classical problem in fluvial hydraulics. It also represents a class of flow problems in coastal engineering such as flow around an island and hence it is a proper test. A cylinder with radius of 0.11 m situated in the centre of a channel is considered here, which is the same as Test 1 used by Yulistiyanto et al. in the numerical and experimental investigations [57]. The channel is 4 m long and 2 m wide. The discharge is $Q = 0.248$ m^3/s; outflow depth is $h_o = 0.185$ m; the bed slope is $\partial z_b/\partial x = -6.25 \times 10^{-4}$ in the flow direction; and the Manning's coefficient is $n_b = 0.012$.

In the numerical simulation, 600×300 square lattices were used. $\Delta x = \Delta y = 0.00667$ m, $\Delta t = 0.00145$ s and $\tau = 1.982$. No-slip boundary condition is used along the wall of the cylinder; slip boundary condition is used for the side walls; the depth at the outflow is set to $h = h_o$; zero gradient for the depth is specified at the inflow boundary; the inflow velocity u is decided based on the discharge; and $v = 0$ at the inflow boundary. After 40,000 steps, a steady state solution was reached.

Comparison of the depths along the centerline of the channel between the computation and the experimental data is shown in Fig. 7.6. The calculation

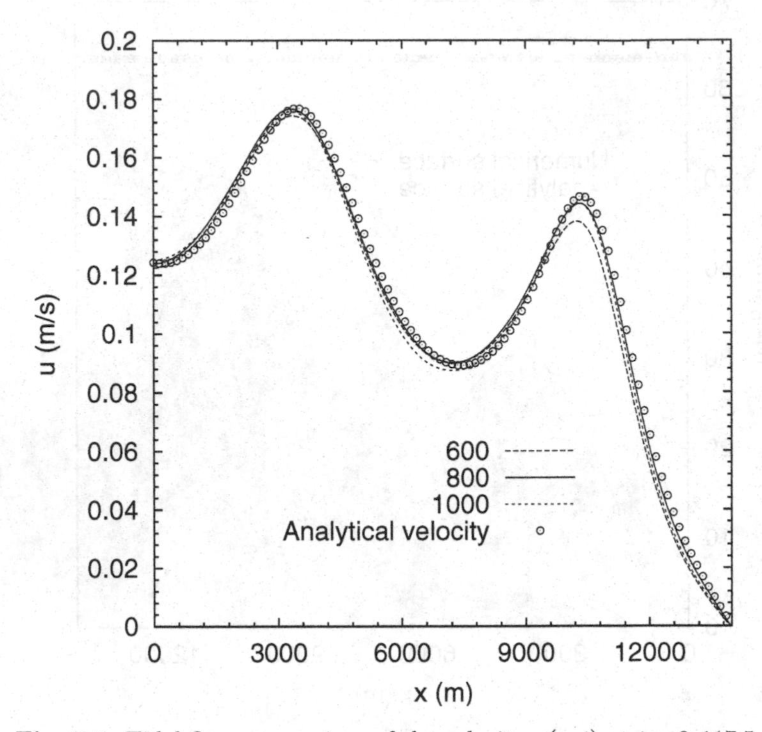

Fig. 7.5. Tidal flow: comparison of the velocity $u(x,t)$ at $t = 9,117.5$ s.

indicates that the maximum relative error for the depth is 9.2%; hence agreement is reasonably good. The contours of the water depths are similar to the pattern observed in the experiment. Fig. 7.7 shows the flow characteristics in the vicinity of the cylinder, which is again similar to Fig. 7.8 given by Yulistiyanto et al. [57] with a different numerical method. In fact, there is a relation between the wake length L_w and the cylinder diameter D, i.e. $L_w \approx 1.3D$ (see Fig. 7.8 for definitions). This is also in good agreement with that obtained by Yulistiyanto et al. [57].

7.3 Flows with Force Term

In this section, we described the numerical results taken from Reference [18]. The main purpose is to assess the centred scheme for force terms in the lattice Boltzmann equation by solving the four benchmark flow problems: a stationary case, a steady flow, a tidal flow and a 2D steady flow. Since the force term related to the bed slope has rich properties as discussed in Section 4.5.1, e.g. it can be a either constant, or linear, or non-linear function, solutions for the shallow water equations with such force terms will be suitable and severe

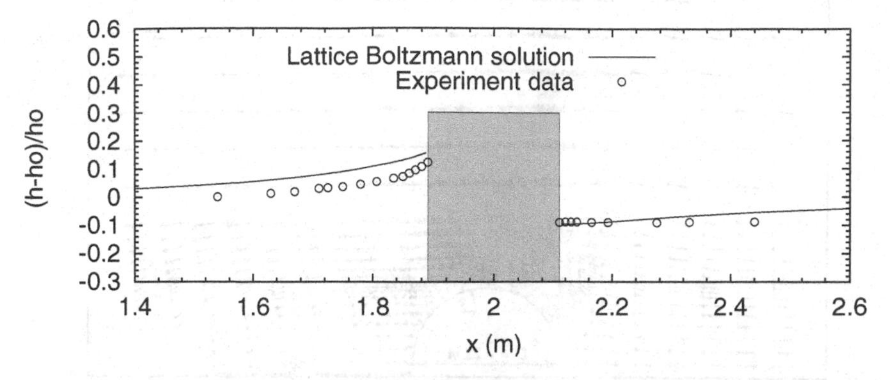

Fig. 7.6. Flow around cylinder: comparison of the depth along channel centerline.

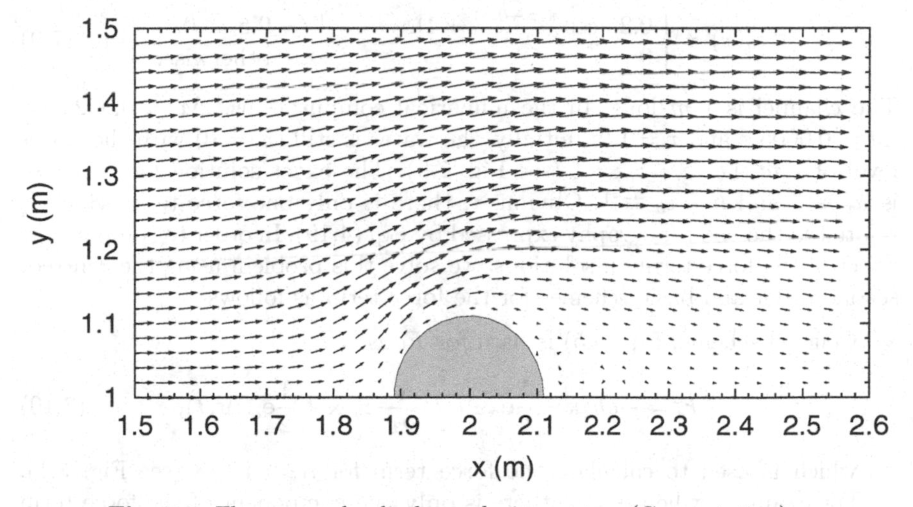

Fig. 7.7. Flow around cylinder: velocity vectors (Computation).

tests for the centred scheme. Therefore, the force term associated with a bed slope is taken into account in all the following numerical tests to demonstrate the accuracy and capability of the centred scheme. Numerical predictions are compared with analytical solutions.

7.3.1 Stationary Case

The first test problem is a still flow over a channel in order to numerically demonstrate that the centred scheme satisfies the $\mathcal{N}$-property. This is a one dimensional problem. The bed topography is the same as that used by LeVeque [58] and is given by

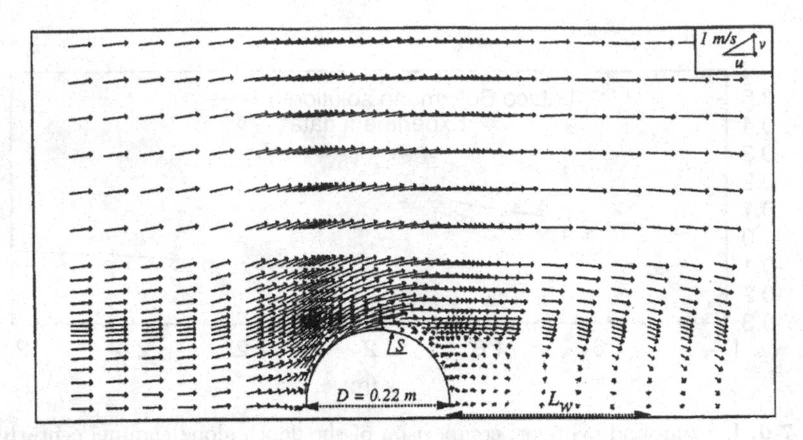

Fig. 7.8. Flow around cylinder: velocity vectors from Ref. [57].

$$z_b(x) = \begin{cases} 0.25[\cos \frac{\pi(x-0.5)}{0.1} + 1], & |x - 0.5| < 0.1, \\ 0, & \text{otherwise.} \end{cases} \tag{7.9}$$

The channel is 1 m long. In the numerical computations, $\Delta x = 0.0025\ m$, $\Delta t = 0.0005\ s$ and $\tau = 1.2$. Initially, the water is still ($u_i = 0$) with the water level of 1 m, i.e. $h + z_b = 1$ (see Fig. 7.9). The exact solution for this case is $u_i \equiv 0$ and $h + z_b \equiv 1$. Obviously, there is only one force term which is related to the bed topography expressed by Eq. (4.19). In order to compare the effect of the force term on solutions, we solve this problem using the centred, second-order and basic schemes for the force term as follows.

- Centred scheme: Eq. (4.5) is used for F_i as

$$F_i = -gh(\mathbf{x} + \frac{1}{2}\mathbf{e}_\alpha \Delta t, t)\frac{\partial}{\partial x_i} z_b(\mathbf{x} + \frac{1}{2}\mathbf{e}_\alpha \Delta t, t), \tag{7.10}$$

which is used to calculate the force term for $\alpha = 1 - 8$ (see Fig. 3.1). For example, when $\alpha = 1$, there is only one component of the force term in x direction F_x because $e_{1y} = 0$. Eq. (7.10) for F_x takes the following discretized form,

$$F_x = -g\frac{[h(x,y,t) + h(x+\Delta x, y, t)]}{2}\frac{[z_b(x+\Delta x, y, t) - z_b(x,y,t)]}{\Delta x}, \tag{7.11}$$

and the similar expressions can be obtained for $\alpha = 2 - 8$.
- Basic scheme: Eq. (4.1) is used, i.e.

$$F_i = -gh(\mathbf{x}, t)\frac{\partial}{\partial x_i} z_b(\mathbf{x}, t), \tag{7.12}$$

which, for $\alpha = 1$, is discretized in the following form of

$$F_x = -gh(x,y,t)\frac{[z_b(x+\Delta x, y, t) - z_b(x - \Delta x, y, t)]}{2\Delta x}. \tag{7.13}$$

- Second-order scheme: Eq. (4.3) is used as

$$F_i = -\frac{1}{2}g\left[h(\mathbf{x},t)\frac{\partial}{\partial x_i}z_b(\mathbf{x},t) + h(\mathbf{x}+\mathbf{e}_\alpha\Delta t,t)\frac{\partial}{\partial x_i}z_b(\mathbf{x}+\mathbf{e}_\alpha\Delta t,t)\right], \quad (7.14)$$

which, for $\alpha = 1$, has the following discretized form,

$$\begin{aligned} F_x = &-\frac{1}{2}gh(x,y,t)\frac{[z_b(x+\Delta x,y,t) - z_b(x-\Delta x,y,t)]}{2\Delta x} \\ &-\frac{1}{2}gh(x+\Delta x,y,t)\frac{[z_b(x+2\Delta x,y,t) - z_b(x,y,t)]}{2\Delta x}. \end{aligned} \quad (7.15)$$

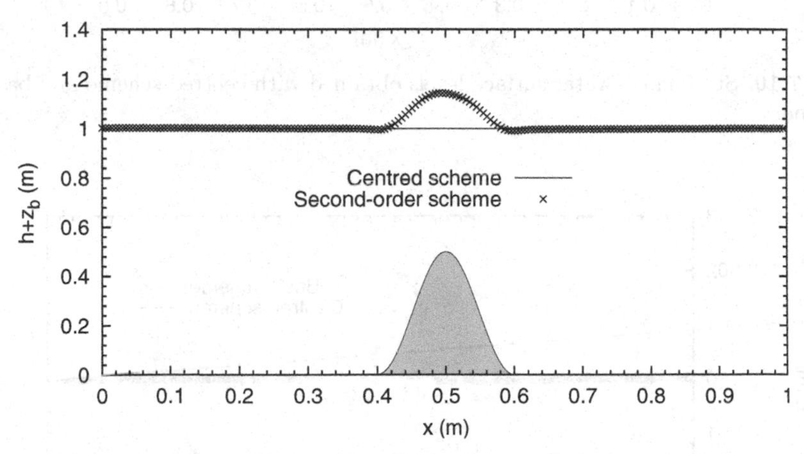

Fig. 7.9. Stationary: water surface levels obtained with centred scheme and second-order scheme.

For the centred scheme, the steady-state solution accurate to within the truncation error of the computer was reached after only 40 iterations; but for basic scheme, it took 7,000 iterations to reach the steady state; and for the second-order scheme it requires as many as 20,000 iterations for the steady-state solution. Comparisons of the water surface levels are shown in Figs. 7.9 and 7.10. Clearly, the centred scheme produces the accurate solution, while the second-order scheme generate solution with relative error as large as about 20%. Although the basic scheme can produce good solution with small error for water surface level, it generates artificial velocities with maximum value of 0.002 m/s as Fig. 7.11 shows. Comparison of velocities between the centred scheme and the second-order scheme is depicted in Fig. 7.12, indicating that the second-order scheme produces as large as 0.04 m/s artificial velocity. These results confirm that only the centred scheme produces accurate solution. This numerically proves that the centred scheme satisfies the $\mathcal{N}$-property, demonstrating the importance of a scheme satisfying the $\mathcal{N}$-property.

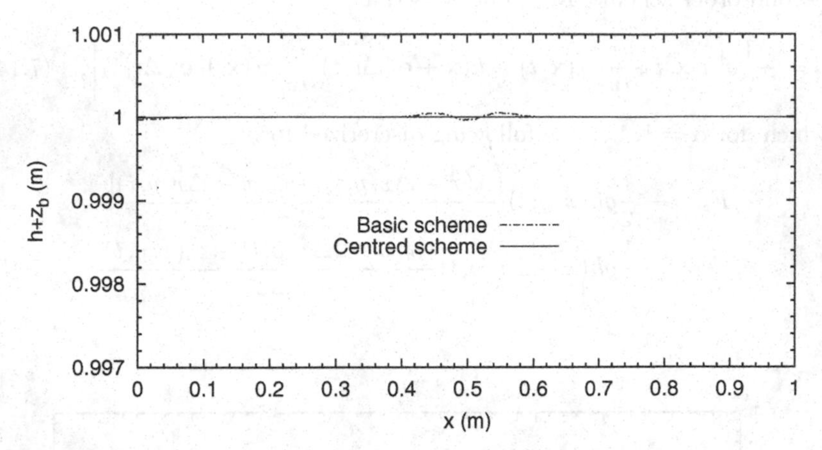

Fig. 7.10. Stationary: water surface levels obtained with centred scheme and basic scheme.

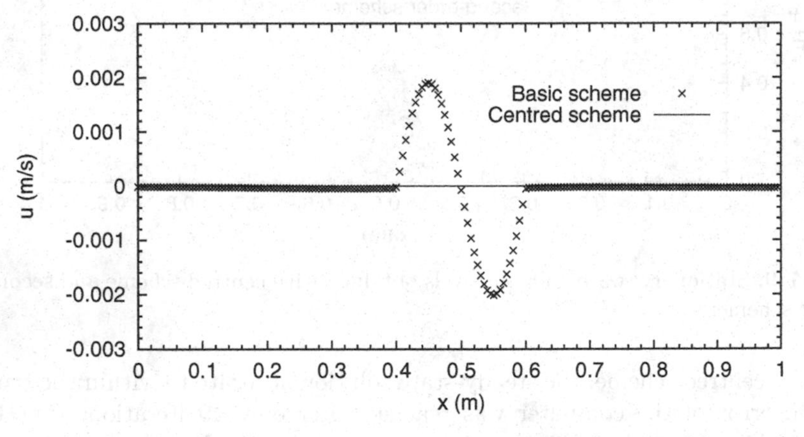

Fig. 7.11. Stationary: velocities obtained with centred scheme and basic scheme.

7.3.2 Steady Flow over an Irregular Bed

The second test is a 1D steady flow over an irregular bed which is the same as that used in Reference [34]. The bed topography is defined in Table 7.1 and shown in Fig. 7.13. Evidently, the force function related to the bed slope is continuous but not differentiable, providing a further severe test for the centred scheme.

In the numerical computations, the discharge per unit width of $q = 10\ m^2/s$ was imposed at the inflow and $h = 20\ m$ was specified at the downstream end. $\Delta x = 3\ m$, $\Delta t = 0.12\ s$ and $\tau = 0.9$. For comparison,

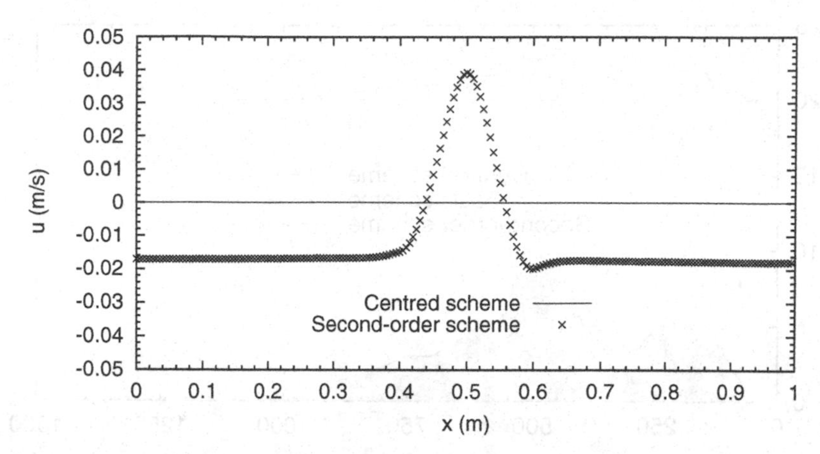

Fig. 7.12. Stationary: velocities obtained with centred scheme and second-order scheme.

Table 7.1. Bed elevation z_b at point x for irregular bed.

x	0	50	100	150	250	300	350	400	425	435	450	475	500	505
z_b	0	0	2.5	5	5	3	5	5	7.5	8	9	9	9.1	9
x	530	550	565	575	600	650	700	750	800	820	900	950	1000	1500
z_b	9	6	5.5	5.5	5	4	3	3	2.3	2	1.2	0.4	0	0

the three schemes were used for the force term. A steady-state solution was reached after 100,000 iterations for the centred scheme; after 200,000 iterations for the basic scheme; and after 200,000 iterations for the second-order scheme but with small time step of $\Delta t = 0.06\ s$ because it was divergent with $\Delta t = 0.12\ s$. Comparison of the water surface levels obtained with the different schemes are shown in Fig. 7.13, from which it is seen that the results for both centred and basic schemes are similar but the second-order scheme generates unphysical oscillations. In order to examine the conservative property, the numerical discharges are depicted in Fig. 7.14, which shows that only the centred scheme retains the constant discharge of $q = 10\ m^2/s$, while both the basic and the second-order schemes produce discharges with a maximum relative error of over 10%. This again confirms that the centred scheme is accurate and conservative.

7.3.3 Tidal Flow over an Irregular Bed

Although a tidal flow over a regular bed as a basic test is presented in Section 7.2.2, we consider a tidal flow over an irregular bed as a further test for the capability of the LABSWE. The bed is the same as that defined in Table 7.1. The initial and boundary conditions are

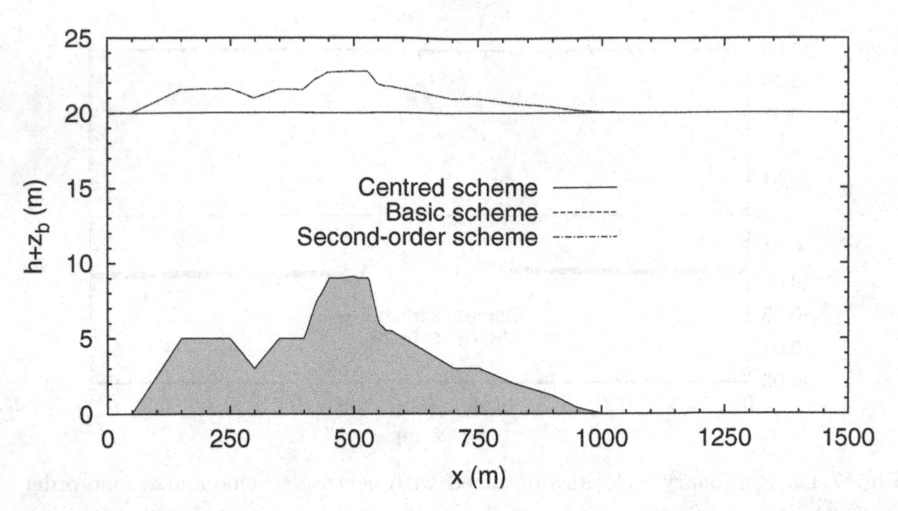

Fig. 7.13. 1D steady: comparison of water surface levels calculated with different schemes.

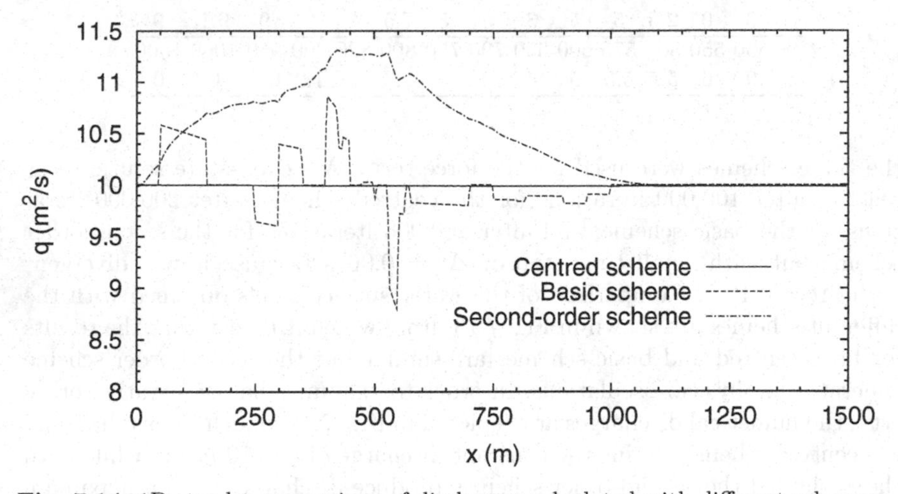

Fig. 7.14. 1D steady: comparison of discharges calculated with different schemes.

$$h(x,0) = 16 - z_b(x), \tag{7.16}$$

$$u(x,0) = 0 \tag{7.17}$$

and

$$h(0,t) = 20 - 4\sin\left[\pi\left(\frac{4t}{86,400} + \frac{1}{2}\right)\right], \tag{7.18}$$

$$u(L,t) = 0, \tag{7.19}$$

where $L = 1,500\ m$ is the channel length.

As indicated in Section 7.2.2, under these conditions, the tidal flow is relatively short and an asymptotic analytical solution is [56],

$$h(x,t) = 20 - z_b(x) - 4\sin\left[\pi\left(\frac{4t}{86,400} + \frac{1}{2}\right)\right], \tag{7.20}$$

and

$$u(x,t) = \frac{(x-L)\pi}{5,400h(x,t)}\cos\left[\pi\left(\frac{4t}{86,400} + \frac{1}{2}\right)\right]. \tag{7.21}$$

In the numerical computations, $\Delta x = 7.5\ m$, $\Delta t = 0.3\ s$ and $\tau = 1.5$. The centred scheme is used for the force term. In order to compare the numerical results with the asymptotic analytical solution, we choose two results at $t = 10,800\ s$ and $t = 32,400\ s$, which correspond to the half-risen tidal flow with maximum positive velocities and to the half-ebb tidal flow with maximum negative velocities. Fig. 7.15 shows a comparison between the predicted water surface and the analytical solution at $t = 10,800\ s$. Comparisons of velocities are depicted in Figs. 7.16 and 7.17. These figures show that there are excellent agreements between the numerical predictions and the analytical solutions. This confirms that the centred scheme is also accurate and conservative for tidal flow over an irregular bed. The results obtained with the basic and the second-order schemes were also carried out. Comparisons of the water surface and the maximum positive velocities at $t = 10,800\ s$ are shown in Figs. 7.18 and 7.19, respectively. It is clearly seen from the figures that only the centred scheme can produce the accurate solution.

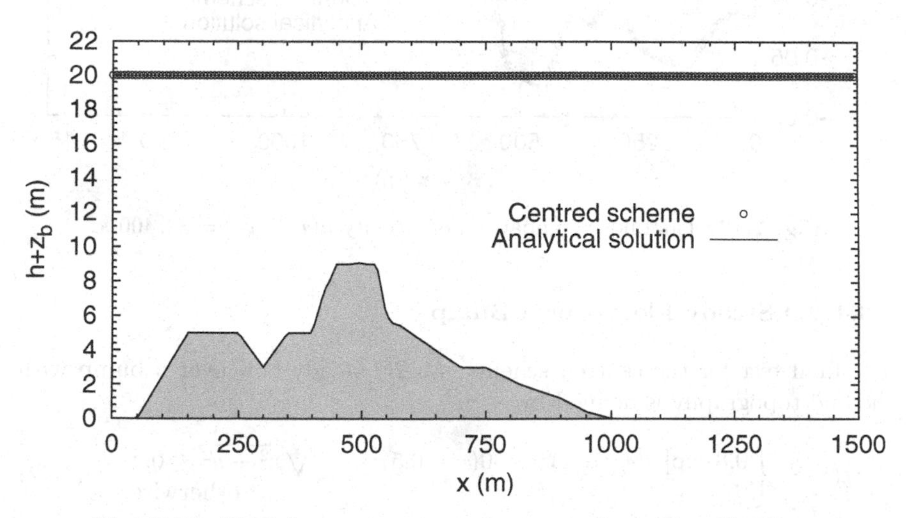

Fig. 7.15. Tidal flow: comparison of water surface at $t = 10,800\ s$.

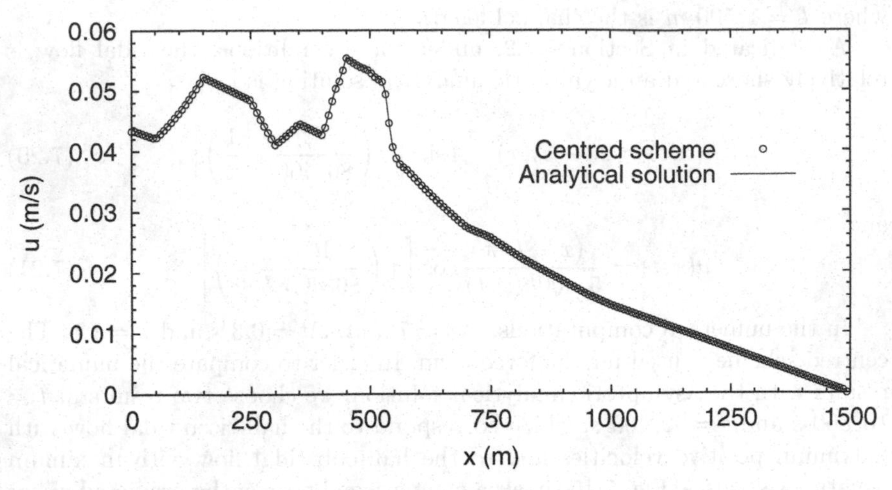

Fig. 7.16. Tidal flow: comparison of velocity $u(x,t)$ at $t = 10,800$ s.

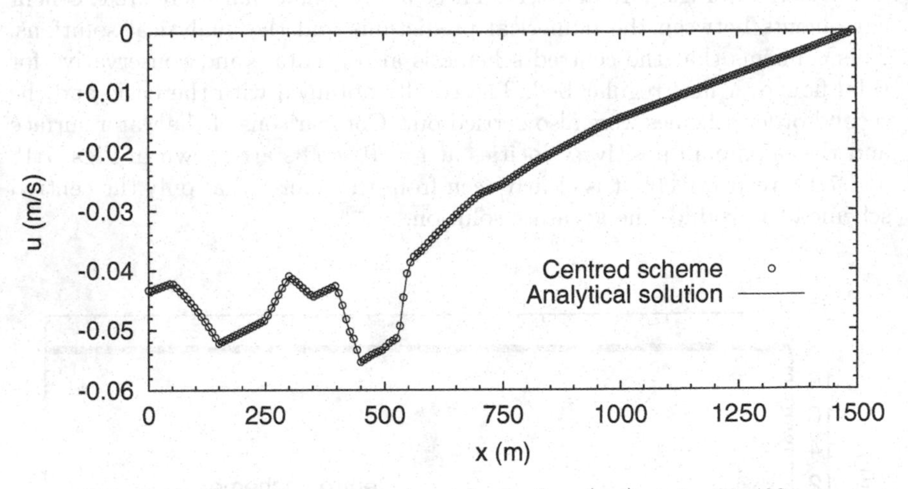

Fig. 7.17. Tidal flow: comparison of velocity $u(x,t)$ at $t = 32,400$ s.

7.3.4 2D Steady Flow over a Bump

The final test for the centred scheme is a 2D steady flow over a bump with the bed topography is defined by

$$z_b(x) = \begin{cases} 0.8\exp[-5(x-1)^2 - 50(y-0.5)^2], & \sqrt{x^2+y^2} \le 0.4, \\ 0, & \text{otherwise}, \end{cases} \tag{7.22}$$

which is shown in Fig. 7.20. The channel is 2 m long and 1 m wide. The discharge is $Q = 0.9$ m^3/s; outflow depth is $h = 1.8$ m. In the computation,

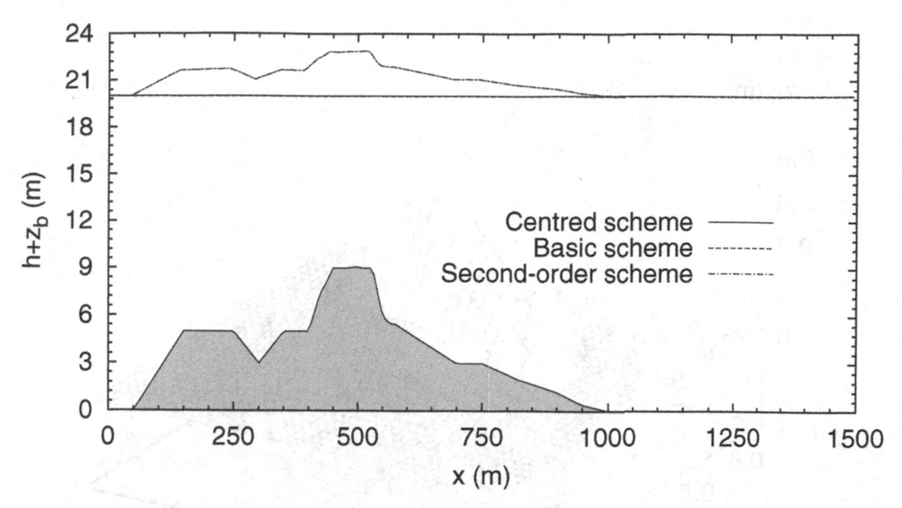

Fig. 7.18. Tidal flow: comparison of water surface levels obtained with different schemes.

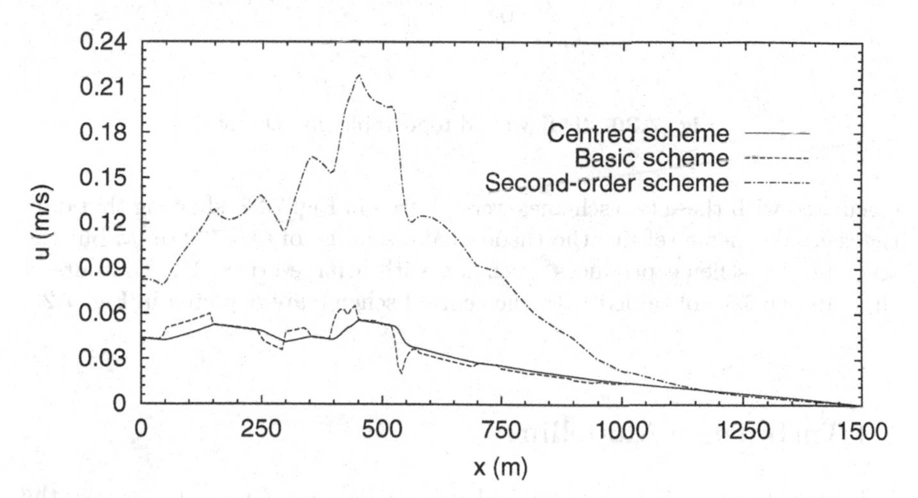

Fig. 7.19. Tidal flow: comparison of velocities calculated with different schemes.

200×100 square lattices were used. $\Delta x = \Delta y = 0.01\ m$, $\Delta t = 0.005\ s$ and $\tau = 1.1$. A steady-state solution was reached after 3,000 iterations for the centred scheme and 6,000 iterations for the second-order scheme. Comparisons of water surface along the channel central line is shown in Fig. 7.21, indicating that there is much difference in the water surface between the two schemes. According to the hydraulics, there is a drop in water surface when water flows over a bump if the flow is subcritical. This is the case presented here and hence the centred scheme provides a correct solution. Comparison of the discharges

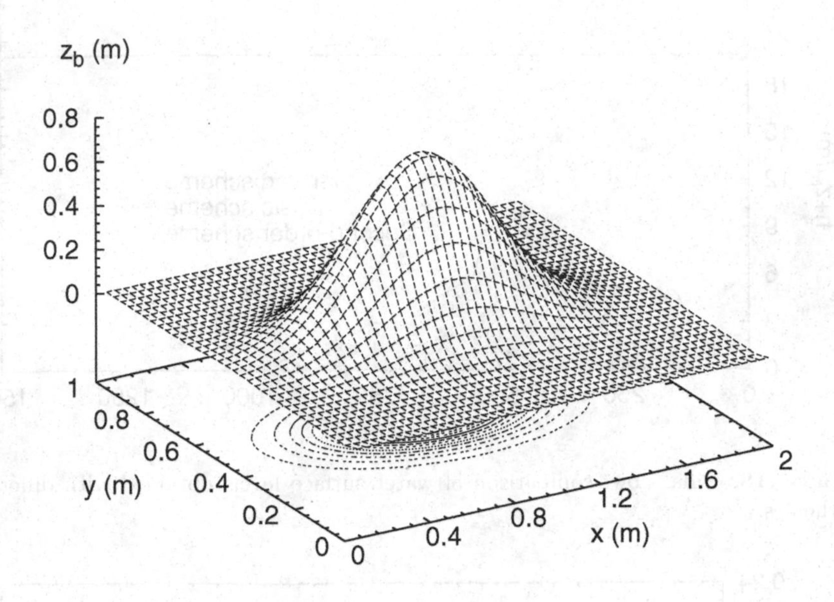

Fig. 7.20. 2D flow: bed topography in 3D plot.

calculated with these two schemes were plotted in Fig. 7.22, showing that only the centred scheme retains the theoretical discharge of $Q = 0.9\ m^3/s$ but the second-order scheme produces discharge with a large error. The contours of the water surface obtained with the centred scheme are depicted in Fig. 7.23.

7.4 Turbulence Modelling

This section presents the numerical calculations by Zhou [16] to test the LABSWE^{TM} for turbulence modelling, i.e. basic characteristics of flows turbulence in a straight channel and a flow over a submerged island. The results were compared with analytical solutions and available experimental data.

7.4.1 Flow in a Straight Channel

Although a flow in a straight channel is a simple case, the basic characteristics of flow turbulence is apparent. Furthermore, there is an analytical solution for the velocity distribution across the channel which can be used to assess the capability of the LABSWE^{TM}.

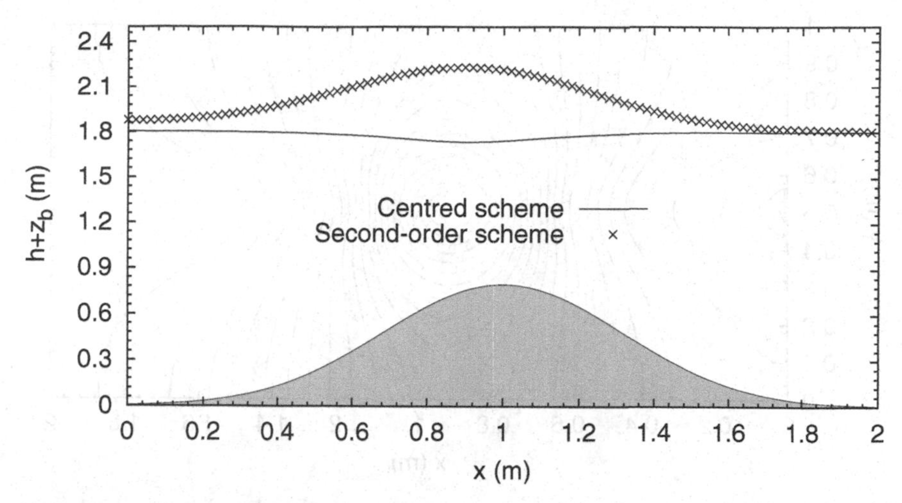

Fig. 7.21. 2D flow: comparison of water surface levels along the channel central line.

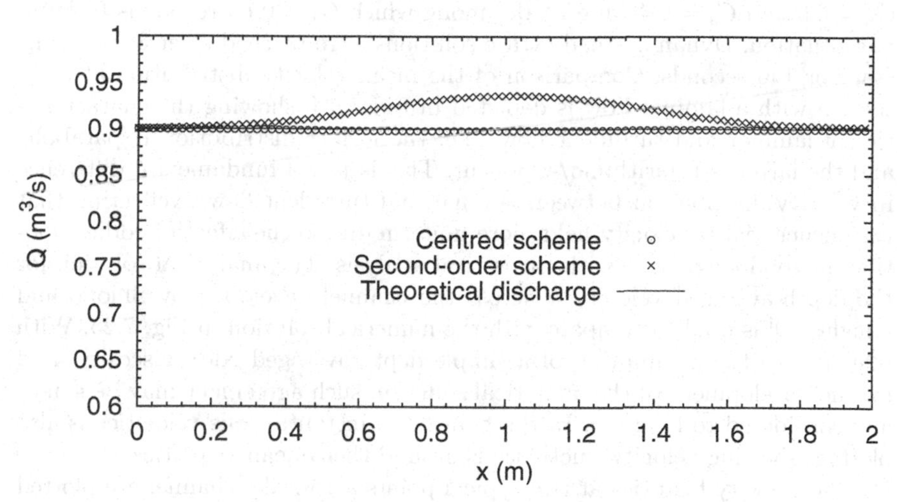

Fig. 7.22. 2D flow: discharges obtained with centred scheme and second-order scheme.

The channel is 8 m long and 0.8 m wide. In the numerical computations, the discharge of $Q = 0.0123\ m^3/s$ was imposed at the inflow and $h = 0.05\ m$ was specified at the downstream end. $\Delta x = 0.02\ m$, $\Delta t = 0.004\ s$ and $\tau = 0.51$. No-slip boundary conditions were used at side walls.

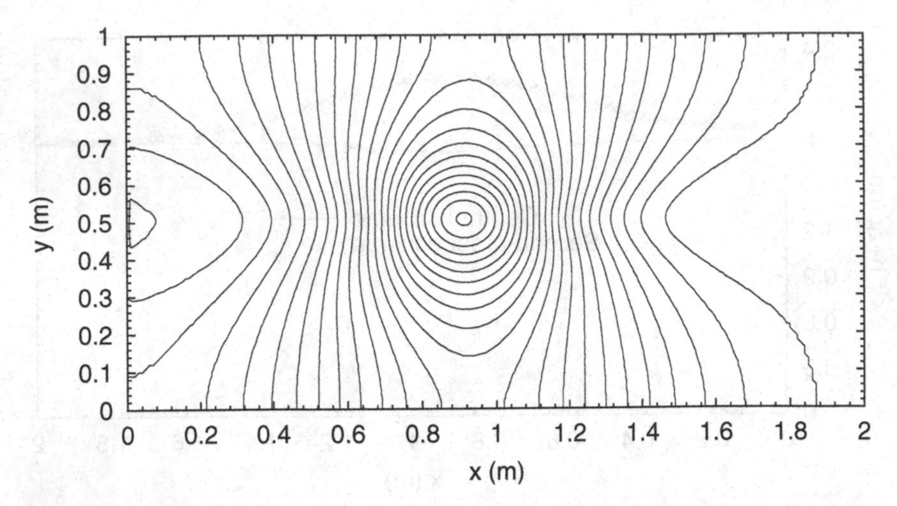

Fig. 7.23. 2D flow: contours of the water surface calculated with centred scheme.

To test the effect of C_s on the solutions, four values, i.e., $C_s = 0$, $C_s = 0.25$, $C_s = 0.3$ and $C_s = 0.4$ were used, among which $C_s = 0$ corresponds to laminar situation. Dynamic steady-state solutions were reached after 35,000 time steps or 140 seconds. Comparison of the mean velocity distribution at cross section with a laminar flow is depicted in Fig. 7.24, showing the characteristics of laminar and turbulent flows, i.e. the former distribution is parabolic and the latter is logarithmic/exponent. This is just a fundamental difference in velocity distribution between laminar and turbulent flows, reflecting that turbulence makes velocity field more uniform due to the effect of eddies' motion, in good agreement with theoretical analysis. The analytical solution for the depth-averaged velocity u across the channel developed by Shiono and Knight [59] is used to compare with the numerical solution in Fig. 7.25. With reference to the assumption of a simple depth-averaged eddy viscosity used for the development of the analytical solution, such agreement may be generally considered to be good. In the figure, the instantaneous velocities is also plotted, showing velocity fluctuations around their mean velocities.

The velocity histories at two typical points across the channel are plotted in Fig. 7.26, showing instantaneous velocities that fluctuate around the mean velocities. Such mean velocities are in fact constants for dynamic steady flows. This confirms that turbulence modelling based on a subgrid-scale stress model can provide more details than time-averaged turbulence model such as k-ϵ model. The vortex contours were shown in Fig. 7.27, from which it can clearly be seen that the closer to side wall, the more eddies, justifying the fact that the wall regions is the source of eddy's generation or the "vortex plant" observed in experimental investigation. Also in the figure shown is the effect of C_s

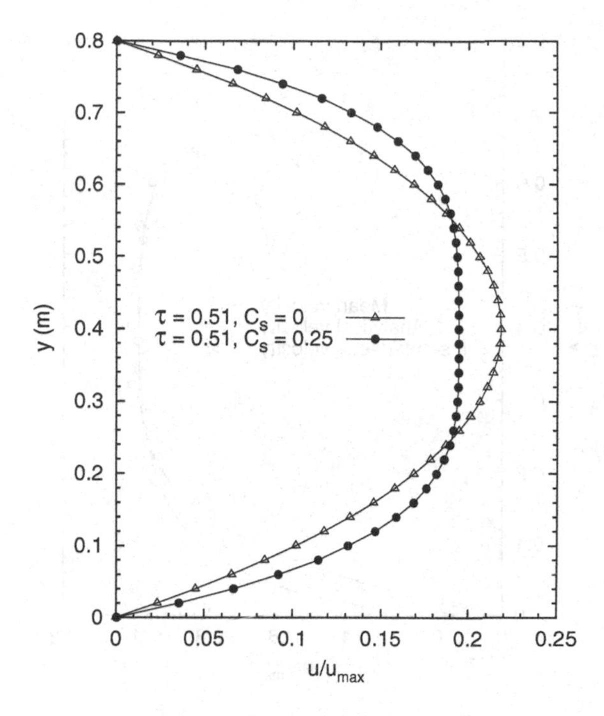

Fig. 7.24. Straight channel: comparison of velocity u distributions.

on the solutions, i.e. the smaller the constant of C_s, the more developed the eddies.

7.4.2 Flow over a Submerged Island

A flow over a submerged island often occurs in natural rivers and seas. The experimental investigation indicates that complex flow phenomenon in which there is vortex shedding happens when the water depth above the island apex is small [60]. This is then a suitable test for numerical methods.

The test used here is the same as Model 4 investigated numerically and experimentally by Lloyd and Stansby [60]. The island shown in Fig. 7.28 is a conical island and consists of the base diameter of 0.75 m and top diameter of 0.05 m with a side slope of 8^0. The channel is 5 m long and 1.52 m wide. The island is situated at $x = 1$ m and $y = 0.76$ m. The longitude bed slope in the channel is 0.000022465, i.e. $\partial z_b/\partial x = -2.2465 \times 10^{-5}$ and the bed friction is neglected.

In the computation, 350 × 106 square lattices were used. $\Delta x = \Delta y = 0.014285$ m, $\Delta t = 0.00357$ s and $\tau = 0.51$. No-slip boundary condition is used along the side walls. The depth at the outflow is set to $h = 0.055$ m; zero gradient for the depth is specified at the inflow boundary; the discharge is

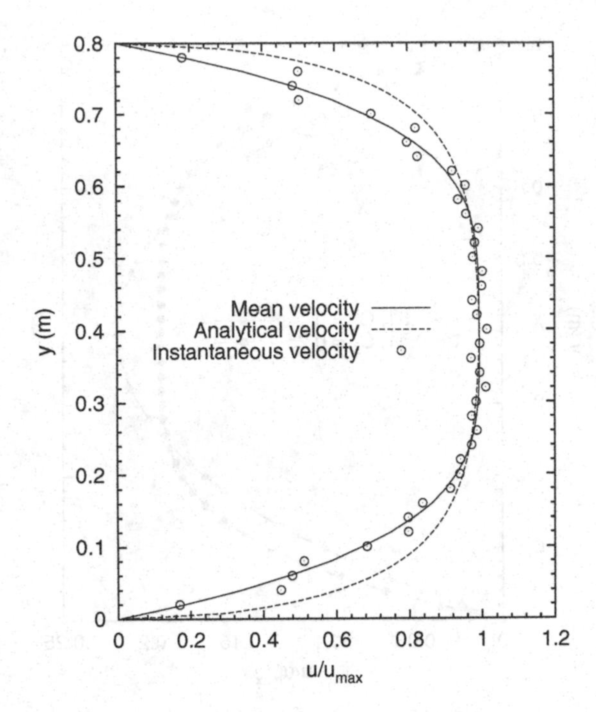

Fig. 7.25. Straight channel: comparison with analytical solution.

$Q = 0.0096\ m^3/s$; and $v = 0$ at the inflow boundary. Several values of C_s were used in the computations.

Comparisons of velocities between the numerical solution and the experimental data were shown in Fig. 7.29, indicating reasonably good agreement at the similar accuracy to that by Lloyd and Stansby [60] with a different numerical method. The velocity vectors are plotted in Fig. 7.30, showing well-developed vortex shedding behind the island, which is similar to that observed in the experiment. Although the velocity vectors between laminar and turbulent flows is not sensibly different for this problem, the plot for vortex contours in Fig. 7.31 exhibits that the eddies are more developed with C_s in the range of $0 < C_s \leq 0.3$ than that with $C_s = 0$ or $C_s \geq 0.4$. This suggests that the LABSWE^{TM} is capable of simulating complex phenomena occurring in shallow water flows, providing detailed structure of flow turbulence.

7.5 More Examples

In this section, we present three additional numerical examples to demonstrate the LABSWE for complicated flows. These tests were carried by Zhou [17] to validate the elastic-collision scheme for slip and semi-slip boundary conditions,

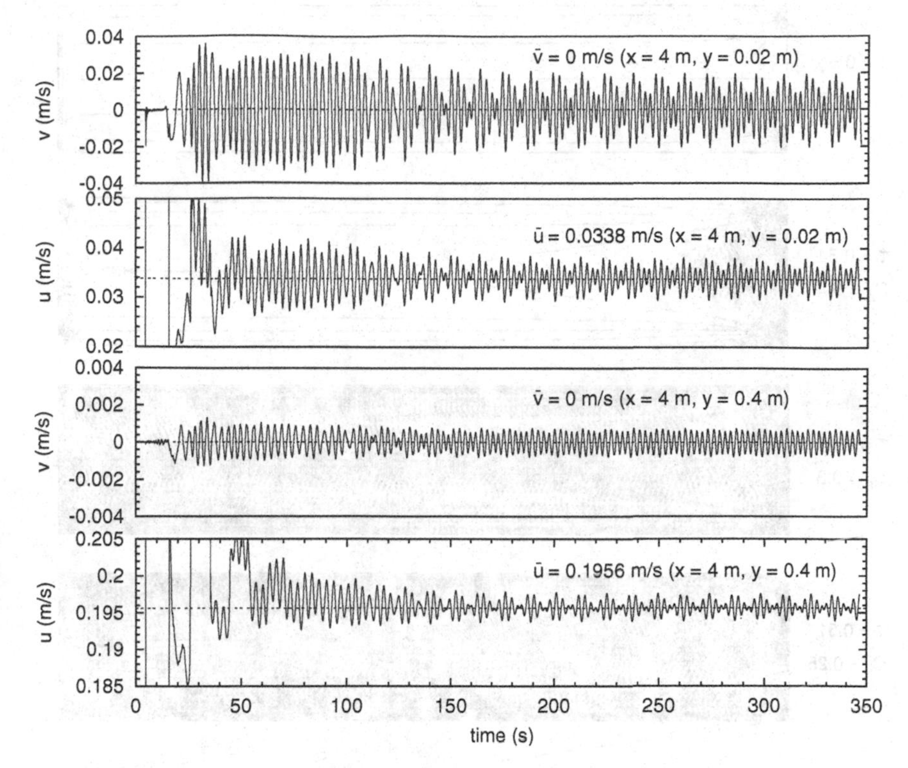

Fig. 7.26. Straight channel: typical velocity history for turbulent flows.

namely a wind-driven circulation in a dish-shaped basin, a flow through a strongly curved channel, and a flow around multiple bodies.

7.5.1 Wind-Driven Circulation in a Dish-Shaped Basin

A wind-driven circulation in lakes is a common flow encountered in nature, which may generate a complex flow phenomenon depending on the bed topography of a lake. Here we consider a uniform wind shear stress applied to the still shallow water in a circular basin. The bed topography is defined with the still water depth given by

$$h = \frac{1}{1.3}\left(\frac{1}{2} + \sqrt{\frac{1}{2} - \frac{1}{2}\frac{r}{R_0}}\right)\ m, \tag{7.23}$$

where r is the distance from the centre of the basin and $R_0 = 193.2\ m$ is the radius of the basin. From Eq. (7.23), it is easily seen that the water depths become progressively deeper towards the basin centre. The configuration is shown in Fig. 7.32. The same dish-shaped basin is also used by Rogers et al. [61] to test a Godunov-type method.

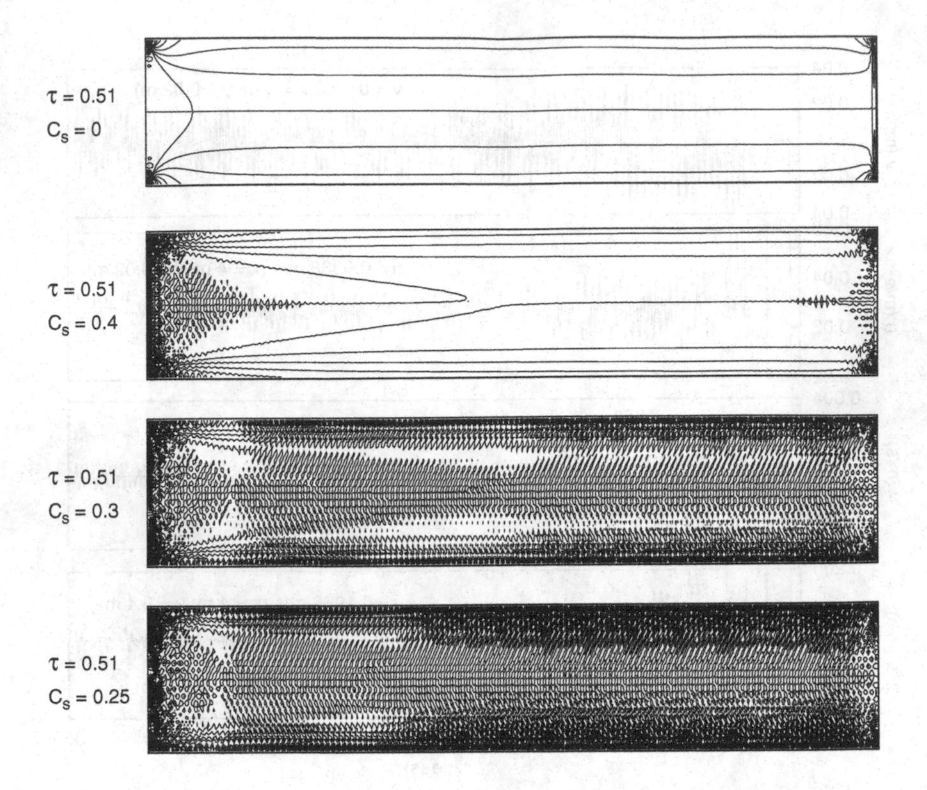

Fig. 7.27. Straight channel: vortex contours and comparisons.

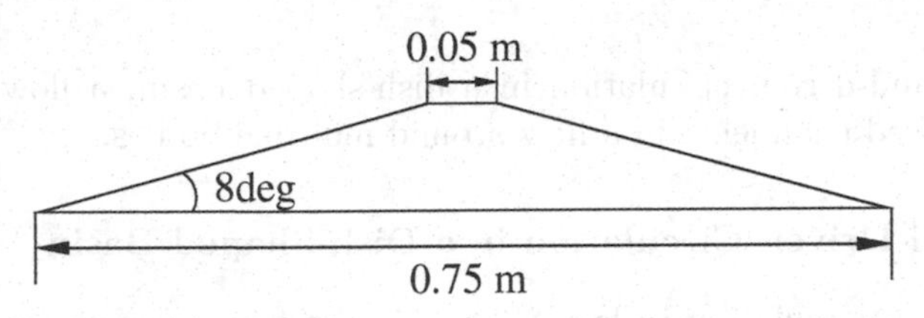

Fig. 7.28. Island: conical model island.

Initially, the water in the basin is still and then a uniform wind shear stress from southwest to northeast is applied (see Fig. 7.33), i.e.

$$\tau_{wx} = 0.004\rho\cos(45^0)\ N/m^2, \qquad \tau_{wy} = 0.004\rho\sin(45^0)\ N/m^2. \tag{7.24}$$

As a result, the steady flow consists of two relatively strong counter-rotating gyres with flow in the deeper water against the direction of the wind. This exhibits complex flow phenomenon involving all types of the boundary treatment described in Section 6.3 for the elastic-collision scheme; hence it is an appropriate test case. An analytical solution for this is developed by Kranenburg [62]. In numerical computation, 200×200 square lattices (see Fig. 7.33) were

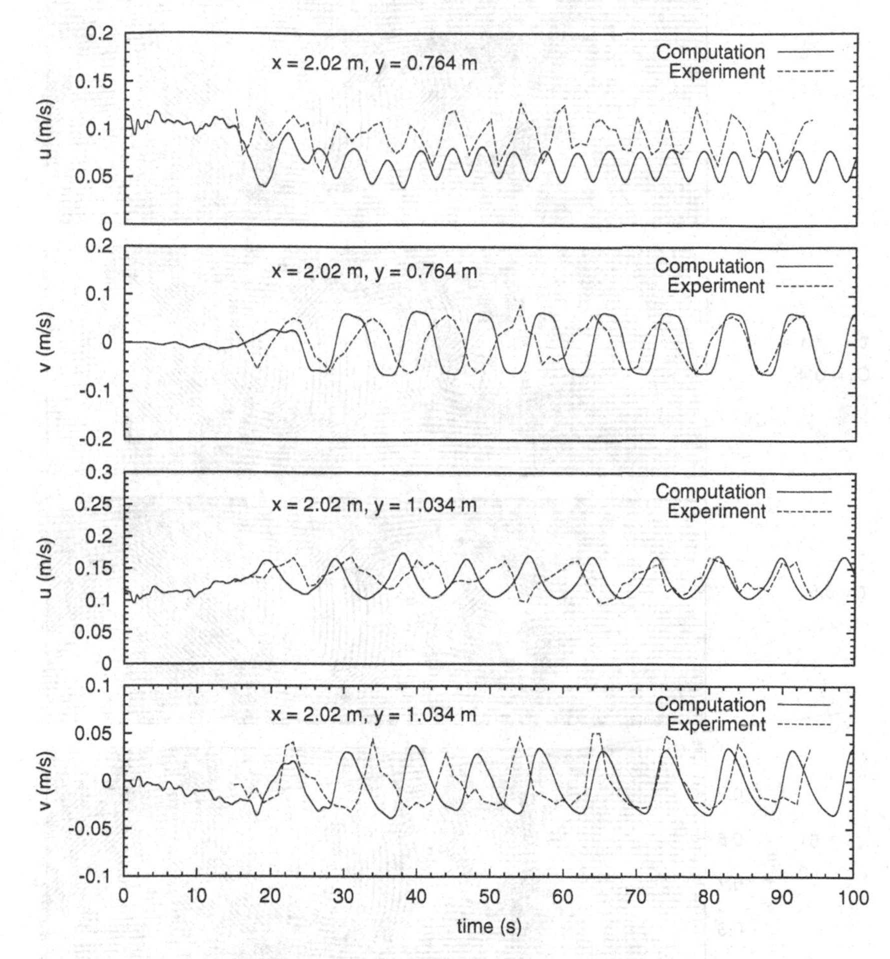

Fig. 7.29. Island: comparison with experimental data ($C_s = 0.3$).

used. $\Delta x = \Delta y = 2\ m$, $\Delta t = 0.2\ s$ and $\tau = 1.325$. Slip boundary conditions such as Eqs. (6.3) - (6.6) together with $C_b = 0$ were used to maintain the same conditions as that of the analytical solution.

The study [17] indicates that slip velocities are effectively independent of the value of the characteristic angle θ_0 if $15^0 \leq \theta_0 \leq 30^0$. In order to compare the numerical results with the analytical solution, we use the results with $\theta_0 = 20^0$ and $\tau = 1.325$. Fig. 7.34 depicts the comparison of the normalized resultant velocity,

$$U = \text{sign}(u+v)\frac{\kappa\sqrt{u^2+v^2}}{u_* \ln Z}, \tag{7.25}$$

perpendicular to the wind direction across the center of the basin, with the analytical solution, where $\kappa = 0.4$ is von Kámán constant, u and v are the

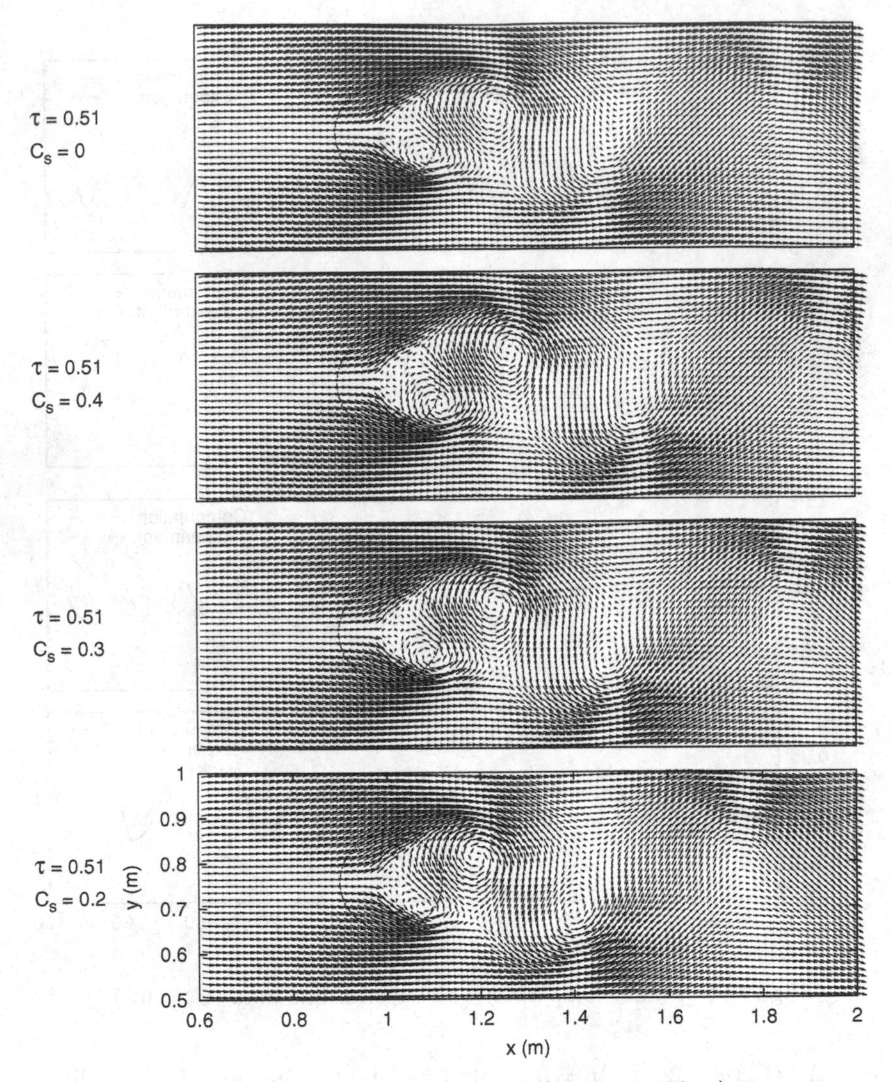

Fig. 7.30. Island: velocity vectors (vortex shedding).

velocity components in the x and y directions respectively, $u_* = \sqrt{\tau_w/\rho}$ is the shear velocity ($\tau_w = \sqrt{\tau_{wx}^2 + \tau_{wy}^2}$) and $Z = H/z_0$ in which $H = 1/1.3\ m$ is the weighted mean water depth and $z_0 = 0.0028\ m$ is the bed roughness height, corresponding to a hydraulically smooth bed. As shown in the figure, the centres of the gyres are well predicted. However, there is discrepancy between numerical predictions and analytical solutions. This is due to the fact that (1) the assumptions of both the rigid-lid approximation for the water surface and a parabolic distribution for the eddy viscosity were used in Kranenburg's

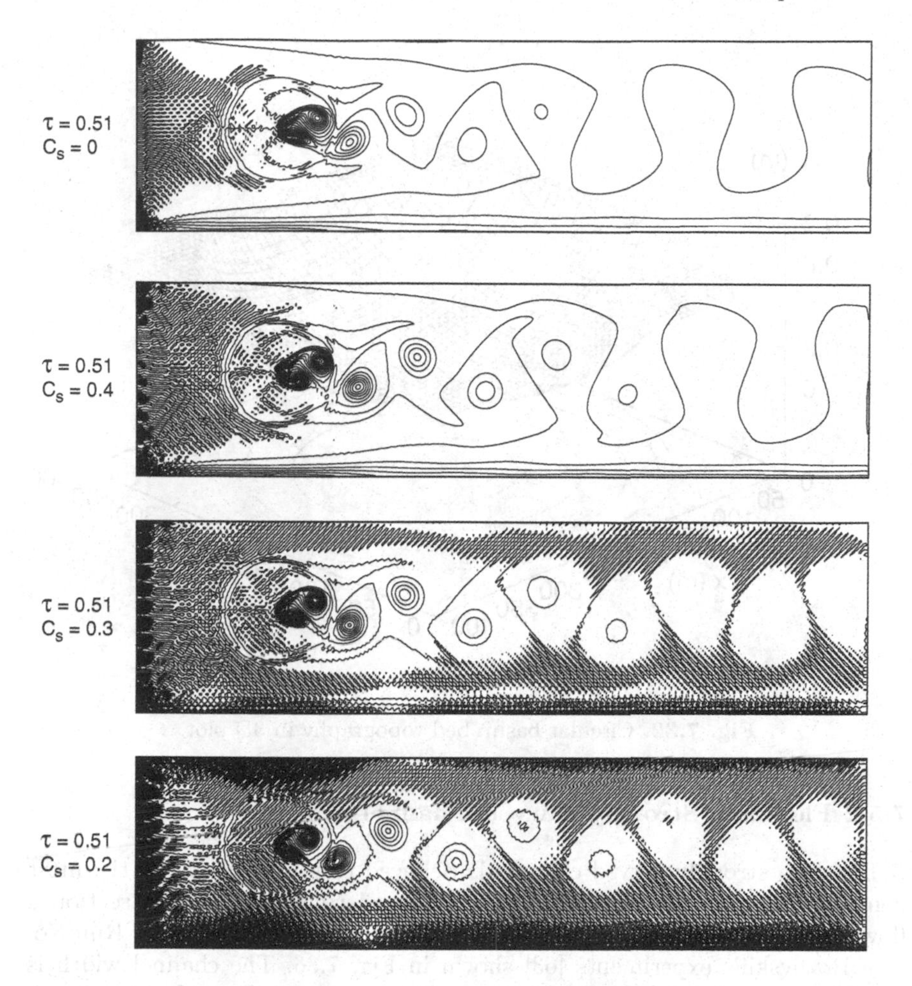

Fig. 7.31. Island: vortex contours.

development for the analytical solution, and (2) the effect of turbulence on flow is not taken into account in the numerical computations. Hence such difference is reasonable and agreement is generally good. The results using the bounce-back scheme are also plotted in Fig. 7.34. Clearly, the bounce-back scheme is not suitable for this problem and provides almost zero velocities at the boundary. The velocity vectors with slip boundary condition are shown in Fig. 7.35, from which two well-developed topographic gyres can be clearly discerned, in agreement with that of the analytical solution. The further detail can be found in Reference [17].

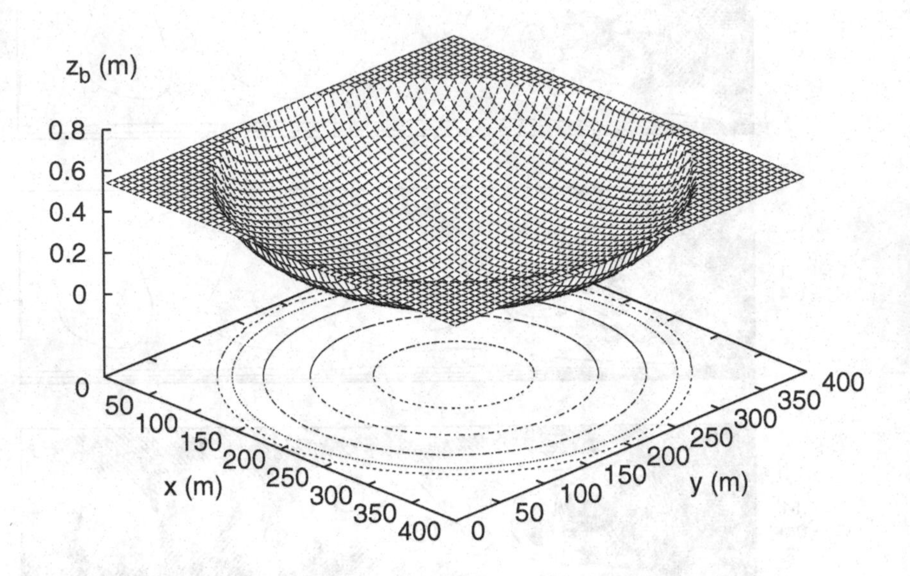

Fig. 7.32. Circular basin: bed topography in 3D plot.

7.5.2 Flow in a Strongly Curved Channel

A flow in a strongly curved channel is believed to represent one of the most complex flows encountered in a natural meandering river. In this section a flow in a 180^0 bend channel is simulated. The channel is the same as Run No. 8 of Rozovskii's experiments [63] shown in Fig. 7.36. The channel width is 0.8 m; the internal radius is 0.4 m; and there is no bed slope in the channel. The flow conditions are: (1) flow discharge is 0.0123 m^3/s; (2) entrance depth is 0.063 m; and (3) the channel bed is rough with Chezy coefficient $C_z = 32$.

In numerical computation, 400×125 square lattices were used. $\Delta x = \Delta y = 0.02\ m$, $\Delta t = 0.01333\ s$ and $\tau = 0.6$. At the upstream boundary, the gradient of depth in flow direction is set to zero; velocity u is accordingly adjusted to retain the constant discharge; and $v = 0$. At the downstream boundary the depth is specified as 0.05 m and the gradients of velocities are set to zeros. These conditions are transformed to suitable conditions for distribution functions with the method described in Section 6.4. At the channel sides, $C_f = 0.03$ and the semi-slip boundary conditions are used, i.e. Eq. (3.16) is solved with modified force term Eq. (6.11) and slip boundary conditions such as Eqs. (6.3) - (6.6). A steady-state solution was reached after 5000 iterations. Comparisons of the tangential velocities between numerical results and experimental data at several cross sections are depicted in Fig. 7.37, showing good

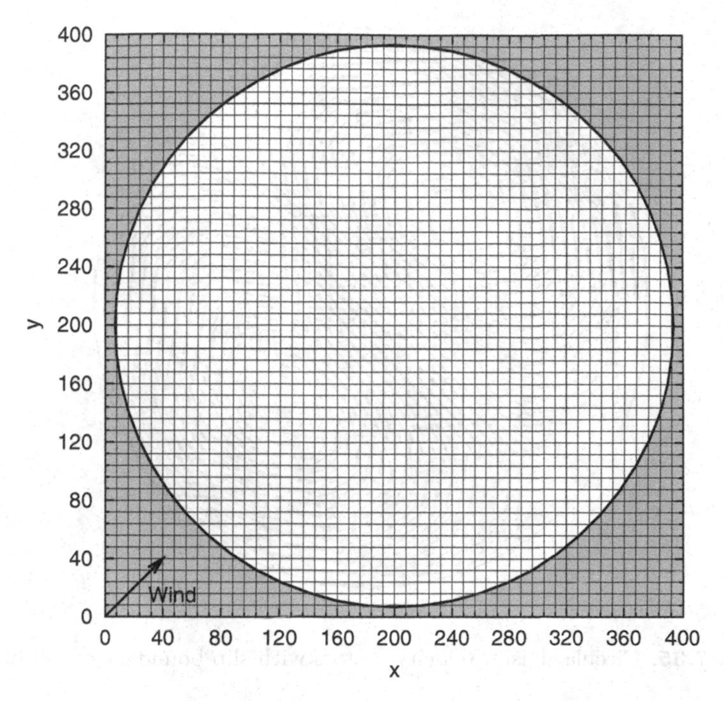

Fig. 7.33. Circular basin: sketch of lattice and basin geometry.

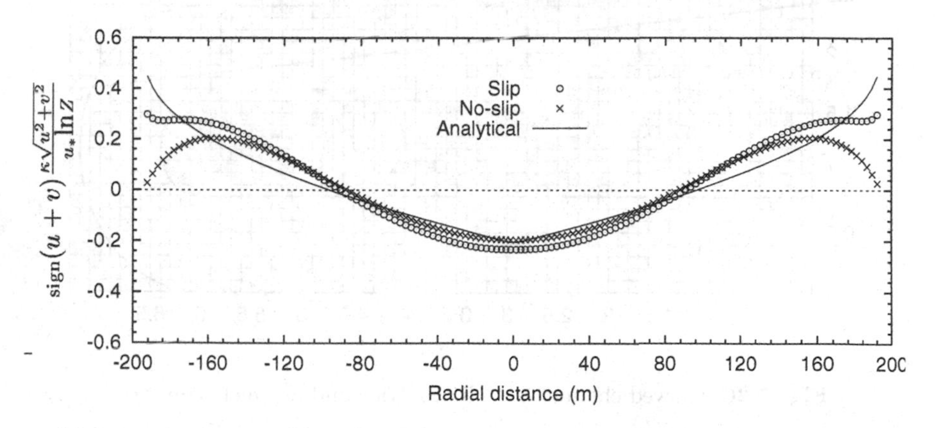

Fig. 7.34. Circular basin: comparison of the normalized resultant velocities.

agreement. The numerical results with no-slip boundary condition at the side boundaries are also plotted in the figure, indicating that the velocities are incorrectly predicted. The velocity vectors are shown in Fig. 7.38.

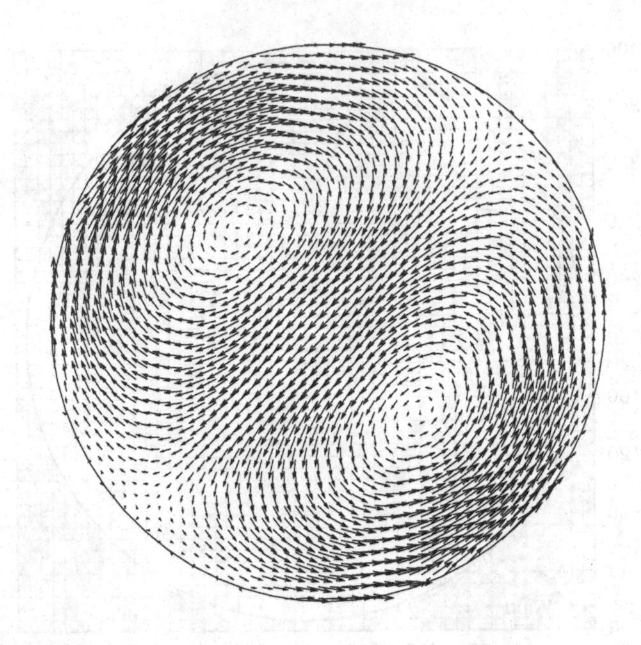

Fig. 7.35. Circular basin: velocity vectors with slip boundary conditions.

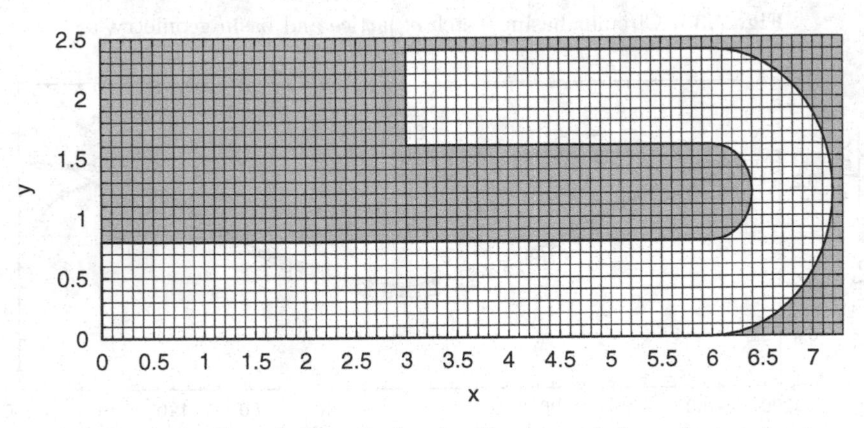

Fig. 7.36. Curved channel: sketch of lattice and channel geometry.

7.5.3 Flow around Multiple Bodies

We carried out a numerical experiment to demonstrate the capability of the elastic-collision scheme for simulating flows around multiple bodies. The body shapes consist of a circle, a square and a triangle. The geometries of the channel and the bodies are shown in Fig. 7.39. The flow discharge is 0.02 m^3/s. The water depth at the downstream is 0.05 m. 600 × 250 lattices were used. $\Delta x = \Delta y = 0.01\ m$, $\Delta t = 0.005\ s$ and $\tau = 0.9$. At the channel sides slip bound-

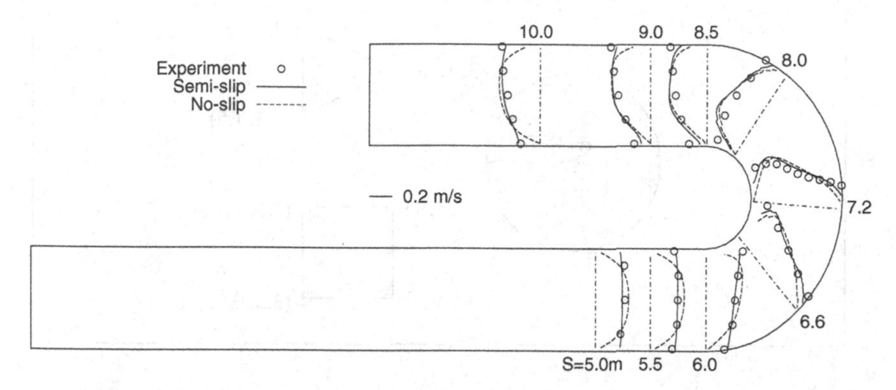

Fig. 7.37. Curved channel: comparisons of the tangential velocities (S is the distance between the cross section and the entrance along the channel central line).

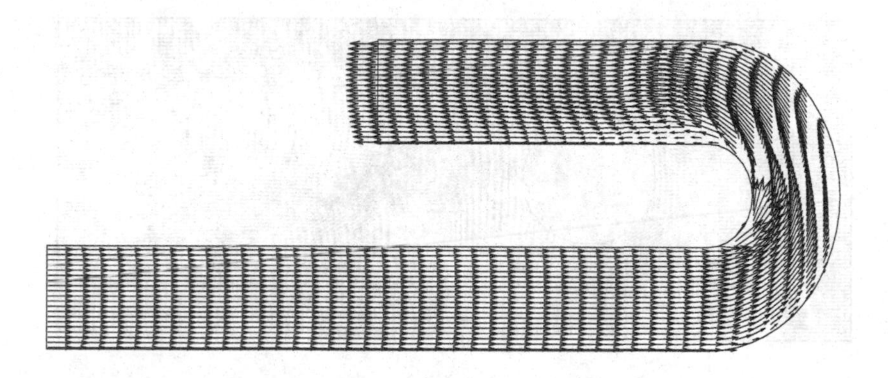

Fig. 7.38. Curved channel: velocity vectors with semi-slip boundary conditions.

ary conditions are used. At the body boundaries semi-slip boundary conditions with $C_f = 0.03$ are used. The velocity vectors are shown in Fig. 7.40. The numerical results with no-slip boundary conditions at the body boundaries are also made. Comparisons of velocity u between the two boundary conditions at the selected cross sections are shown in Fig. 7.41, indicating a significant difference in the velocities in the vicinity of the bodies.

7.6 Closure

In this book, we have described the lattice Boltzmann methods for shallow water flows (LABSWE and LABSWE^{TM}). The illustrated examples have indicated that the methods have many potential capabilities in simulating shallow

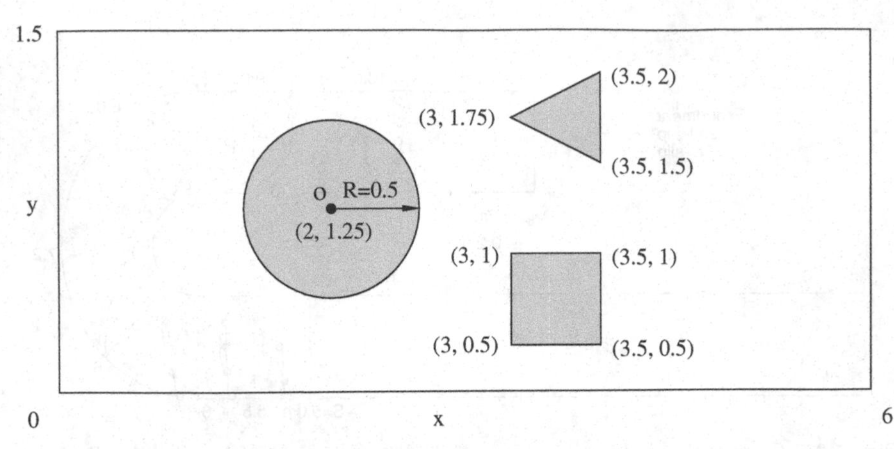

Fig. 7.39. Multiple bodies: geometries of channel and bodies.

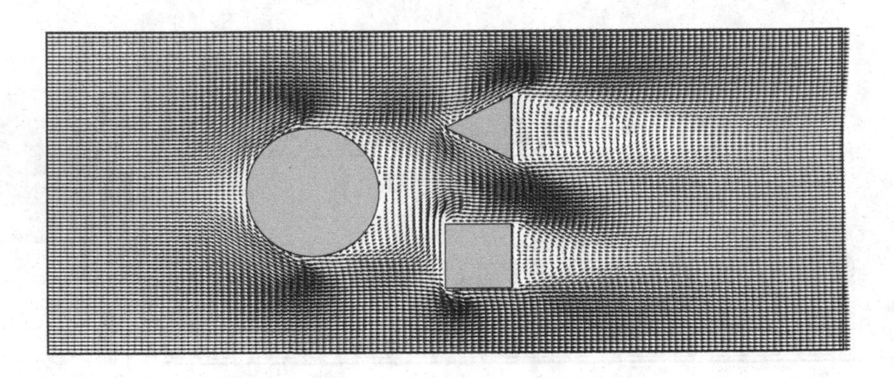

Fig. 7.40. Multiple bodies: velocity vectors with semi-slip boundary conditions.

water flows. The detailed procedures of the methods are addressed in order to enable the readers to write their own practical computer code for different shallow water flow problems. In addition, a sample code is provided in Appendix B which can be used as a starting point for the readers to modify it further and turn it into a practical code.

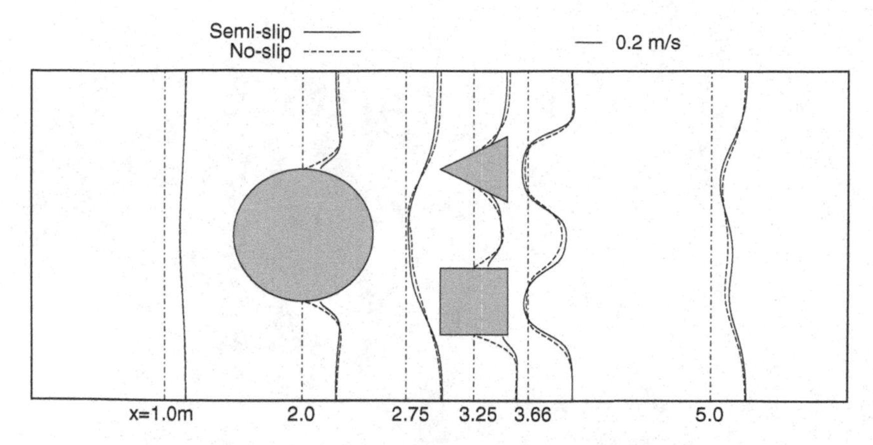

Fig. 7.41. Multiple bodies: comparisons of velocity u.

A

LABSWE on Hexagonal Lattice

A lattice Boltzmann model for the shallow water equations using the 7-speed hexagonal lattice shown in Fig. ?? can be developed in a similar manner to that on the 9-speed square lattice as

$$f_\alpha(\mathbf{x} + \mathbf{e}_\alpha \Delta t, t + \Delta t) - f_\alpha(\mathbf{x}, t) = -\frac{1}{\tau}(f_\alpha - f_\alpha^{eq}) + \frac{\Delta t}{3e^2} e_{\alpha i} F_i(\mathbf{x}, t) \qquad \text{(A.1)}$$

with the velocity vector of particles defined by

$$\mathbf{e}_\alpha = \begin{cases} (0, 0), & \alpha = 0, \\ e\left[\cos(\frac{(\alpha-1)\pi}{3}), \sin(\frac{(\alpha-1)\pi}{3})\right], & \alpha = 1 - 6 \end{cases} \qquad \text{(A.2)}$$

and the local equilibrium function f_α^{eq} expressed as

$$f_\alpha^{eq} = \begin{cases} h - \frac{gh^2}{e^2} + \frac{h}{e^2} u_i u_i, & \alpha = 0, \\ \frac{gh^2}{6e^2} + \frac{h}{3e^2} e_{\alpha i} u_i + \frac{2h}{3e^4} e_{\alpha i} e_{\alpha j} u_i u_j - \frac{h}{2e^2} u_i u_i, & \alpha = 1 - 6. \end{cases} \qquad \text{(A.3)}$$

The physical variables, water depth and velocity, are still given by Eqs. (3.46) and (3.51). For turbulent flows, τ in Eq. (A.1) is simply replaced with the total relaxation time τ_t determined by Eq. (5.18).

B

LABSWE Code

This appendix includes a sample code LABSWE.f90 for LABSWE described in the book and a main.f90 to demonstrate how to use the code to simulate shallow water flows. They are written in FORTRAN-90 language and may be compiled with any FORTRAN-90 compiler. As indicated in the code, it can be served as a starting point both to practice the LABSWE and to turn it into a practical code with the details provided in the book.

B.1 LABSWE Module

This is the core module for the LABSWE written in FORTRAN 90 Language. An example of its application is given in Section B.2.

```
!-----------------------------------------------------------!
!                        LABSWE.f90                         !
!                                                           !
! This module written in FORTRAN 90 implements the lattice  !
! Boltzmann method for shallow water equations (LABSWE),    !
! based on the 9-speed square lattices. It is included as   !
! a sample code in the book and hence only highlights the   !
! major procedure in the LABSWE under conditions that time  !
! step and lattice spacing are taken as units. Also the     !
! module provides periodic boundary conditions and no-slip  !
! boundary conditions in y direction boundaries. It should  !
! be pointed that the module can easily be adapted into a   !
! practical code by changing these settings or adding more  !
! boundary conditions. A sample file main.f90 is also       !
! included to show how to use the module.                   !
!                                                           !
!                          J.G. Zhou, Peterborough, 2003    !
!-----------------------------------------------------------!
```

```
!                        List of Major Variables                       !
!                                                                      !
! a, x, y    - Loop integers                                           !
! ex, ey     - x and y components of particles' velocities              !
! f          - Distribution function                                   !
! feq        - local equilibrium distribution function                 !
! force_x    - x-direction component of force term                     !
! force_y    - y-direction component of force term                     !
! ftemp      - Temple distribution function                            !
! gacl       - Gravitational acceleration                              !
! h          - water depth                                             !
! Lx, Ly     - Total lattice numbers in x and y directions             !
! nu         - Molecular viscosity                                     !
! tau        - Relaxation time                                         !
! u, v       - x and y components of flow velocity                     !
!----------------------------------------------------------------------!
module LABSWE

   implicit none

   integer:: Lx,Ly,x,y,a
   real:: tau,nu,gacl = 9.81
   real, dimension(9):: ex,ey
   real, allocatable, dimension(:,:):: u,v,h,force_x,force_y
   real, allocatable, dimension(:,:,:):: f,feq,ftemp

contains

subroutine setup

   ! declare a local real for the quarter of PI
   real:: quarter_pi

   ! set constant PI
   quarter_pi = atan(1.)

   ! calculate molecular viscosity
   nu = (2.*tau-1)/6.

   ! compute the particle velocities
   do a = 1, 8
      if (mod(a,2) == 0) then
         ex(a) = sqrt(2.)*cos(quarter_pi*real(a-1))
         ey(a) = sqrt(2.)*sin(quarter_pi*real(a-1))
      else
```

```
        ex(a) = cos(quarter_pi*real(a-1))
        ey(a) = sin(quarter_pi*real(a-1))
     end if
  end do
  ex(9) = 0.; ey(9) = 0.

  ! compute the equilibrium distribution function feq
  call compute_feq

  ! Set the initial distribution function to feq
  f = feq

  return
end subroutine setup

subroutine collide_stream

  ! This calculate distribution function with the LABSWE

  ! local working integers
  integer:: xf,yf,xb,yb

  do y = 1, Ly

     yf = y + 1
     yb = y - 1

  do x = 1, Lx

     xf = x + 1
     xb = x - 1

     ! Following 4 lines Implement periodic BCs
     if (xf > Lx) xf = xf - Lx
     if (xb < 1) xb = Lx + xb
     if (yf > Ly) yf = yf - Ly
     if (yb < 1) yb = Ly + yb

     ! start streaming and collision
     ftemp(1,xf,y) = f(1,x,y)-(f(1,x,y)-feq(1,x,y))/tau&
          & + 1./6.*(ex(1)*force_x(x,y)+ey(1)*force_y(x,y))
     ftemp(2,xf,yf) = f(2,x,y)-(f(2,x,y)-feq(2,x,y))/tau&
          & + 1./6.*(ex(2)*force_x(x,y)+ey(2)*force_y(x,y))
     ftemp(3,x,yf) = f(3,x,y)-(f(3,x,y)-feq(3,x,y))/tau&
          & + 1./6.*(ex(3)*force_x(x,y)+ey(3)*force_y(x,y))
```

```
     ftemp(4,xb,yf) = f(4,x,y)-(f(4,x,y)-feq(4,x,y))/tau&
          & + 1./6.*(ex(4)*force_x(x,y)+ey(4)*force_y(x,y))
     ftemp(5,xb,y) = f(5,x,y)-(f(5,x,y)-feq(5,x,y))/tau&
          & + 1./6.*(ex(5)*force_x(x,y)+ey(5)*force_y(x,y))
     ftemp(6,xb,yb) = f(6,x,y)-(f(6,x,y)-feq(6,x,y))/tau&
          & + 1./6.*(ex(6)*force_x(x,y)+ey(6)*force_y(x,y))
     ftemp(7,x,yb) = f(7,x,y)-(f(7,x,y)-feq(7,x,y))/tau&
          & + 1./6.*(ex(7)*force_x(x,y)+ey(7)*force_y(x,y))
     ftemp(8,xf,yb) = f(8,x,y)-(f(8,x,y)-feq(8,x,y))/tau&
          & + 1./6.*(ex(8)*force_x(x,y)+ey(8)*force_y(x,y))
     ftemp(9,x,y) = f(9,x,y) - (f(9,x,y)-feq(9,x,y))/tau

  end do

  end do

  return
end subroutine collide_stream

subroutine solution

  ! compute physical variables h, u and v

  !Set the distribution function f
  f = ftemp

  !compute the velocity and depth
  h = 0.0
  u = 0.0
  v = 0.0
  do a = 1, 9
     h(:,:) = h(:,:) + f(a,:,:)
     u(:,:) = u(:,:) + ex(a)*f(a,:,:)
     v(:,:) = v(:,:) + ey(a)*f(a,:,:)
  end do
  u = u/h
  v = v/h

  return
end subroutine solution

subroutine compute_feq

  ! this compute local equilibrium distribution function
```

```
  do a = 1, 8
     if (mod(a,2) == 0) then
        feq(a,:,:) = gacl*h(:,:)*h(:,:)/24. +&
             & h(:,:)/12.*(ex(a)*u(:,:)+&
             & ey(a)*v(:,:))+h(:,:)/8.&
             & *(ex(a)*u(:,:)*ex(a)*u(:,:)+&
             & 2.*ex(a)*u(:,:)*ey(a)*v(:,:)+&
             & ey(a)*v(:,:)*ey(a)*v(:,:))-&
             & h(:,:)/24.*(u(:,:)*u(:,:)+&
             & v(:,:)*v(:,:))
     else
        feq(a,:,:) = gacl*h(:,:)*h(:,:)/6. +&
             & h(:,:)/3.*(ex(a)*u(:,:)+&
             & ey(a)*v(:,:))+h(:,:)/2.&
             & *(ex(a)*u(:,:)*ex(a)*u(:,:)+&
             & 2.*ex(a)*u(:,:)*ey(a)*v(:,:)+&
             & ey(a)*v(:,:)*ey(a)*v(:,:))-&
             & h(:,:)/6.*(u(:,:)*u(:,:)+&
             & v(:,:)*v(:,:))
     end if
  end do
  feq(9,:,:) = h(:,:)-5.*gacl*h(:,:)*h(:,:)/6.-&
       & 2.*h(:,:)/3.*(u(:,:)**2.+v(:,:)**2.)

  return
end subroutine compute_feq

subroutine Noslip_BC

  ! this is for noslip boundary with Bounce back scheme

  ! for lower boundary
  do a = 2, 4
     ftemp(a,:,1) = ftemp(a+4,:,1)
  end do

  ! for upper boundary
  do a = 6, 8
     ftemp(a,:,Ly) = ftemp(a-4,:,Ly)
  end do

  return
end subroutine Noslip_BC

end module LABSWE
```

B.2 An Example

This section presents a sample program main.f90 to demonstrate how to use the LABSWE module for simulating shallow water flows .

```
!-----------------------------------------------------------!
!                         main.f90                          !
!                                                           !
! The file main.f90 in FORTRAN 90 is included in the book  !
! to show how to use the module LABSWE.f90. Any FORTRAN-90 !
! compiler may be used to compile, e.g.                     !
!          "f90 LABSWE.f90 main.f90 -o labswe".             !
! It simulates a flow in straight channel under constant    !
! force with the periodic boundary conditions in the x      !
! direction and no-slip boundary condition in y direction. !
! Consequently, a steady solution is obtained after 9000th !
! time steps, showing a typical laminar flow where a        !
! parabolic distribution in velocity across the channel is !
! well developed.                                           !
!                                                           !
!                           J.G. Zhou, Peterborough, 2003 !
!-----------------------------------------------------------!
!                 List of Major Variables                   !
!                                                           !
! Fr         - Froude number                                !
! ho         - Initial water depth                          !
! itera_no   - Total iteration number or time steps         !
! time       - Time step or iteration counter               !
! uo, vo     - initial velocities                           !
!-----------------------------------------------------------!
program main

  ! call the module LABSWE
  use LABSWE;  implicit none

  ! declare local working variables
  integer:: time,itera_no
  real:: ho,uo,vo
  character:: fdate*24,td*24 ! get date for output

  ! Total iteration numbers
  itera_no = 9000
```

```
! constants for initializing flow field.
ho = 0.05
uo = 0.25
vo = 0.

! define total lattice numbers in x and y directions
Lx = 100; Ly = 50

! assign a value for the relaxation time
tau = 1.5

! allocate dimensions for dynamic arrays
allocate (f(9,Lx,Ly),feq(9,Lx,Ly),ftemp(9,Lx,Ly),h(Lx,Ly),&
        & force_x(Lx,Ly),force_y(Lx,Ly),u(Lx,Ly),v(Lx,Ly))

! initialize the depth and velocities
h = ho
u = uo
v = vo

! Set constant force
force_x = 0.000024
force_y = 0.

! prepare the calculations
call setup

! main loop for time marching
time = 0
timStep: do

   time = time+1

   ! Streaming and collision steps
   call collide_stream

   ! Noslip BC apply
   call Noslip_BC

   ! Calculate h, u & v
   call solution

   ! update the feq
   call compute_feq
```

```
      write(6,'(I6,A2,3(E16.8,A2))') time,'  ',u(Lx/2,Ly/2)

      if (time == itera_no) exit

   end do timStep

   write(6,*)
   write(6,*)' Writing results in file: result.dat...'
   open(66,file='result.dat',status='unknown')
   td=fdate()
   write(66,*) '# Date: ',td
   write(66,*) '# Fr =',u(1,Ly/2)/sqrt(gacl*h(1,Ly/2))
   write(66,*) '# tau =',tau,',  u0 =',uo
   write(66,*)'# Iteration No.: ',itera_no
   write(66,'(1X,A6,I3,A9,I3)') '# Lx = ', Lx, '  Ly = ', Ly
   write(66,*) '#          Results of the computations'
   write(66,'(1X,A3,A4,A11,2A12)') '# x','y','h(i,j)',&
                                        & 'u(i,j)','v(i,j)'
   write(66,*) '#-------------------------------------------'

   do x = 1, Lx
      do y = 1, Ly
         write(66,'(2i4,3f12.6)')x,y,h(x,y),u(x,y),v(x,y)
      end do
   end do
   close(66)

end program main
```

After the above codes are compiled and run, a flow pattern with vectors shown in Fig. B.1 is obtained.

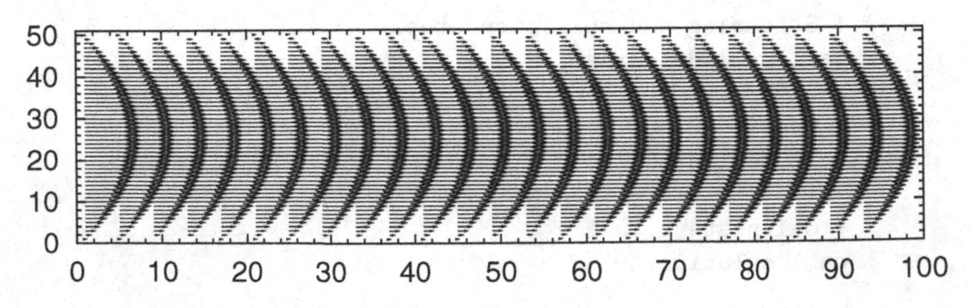

Fig. B.1. Sample code: flow pattern in a straight channel.

References

1. A. Pires, D. P. Landau, and H. Herrmann, editors. *Computational Physics and Cellular Automata.* World Scientific, 1990.
2. B. Chopard and M. Droz. *Cellular Automata Modeling of Physical Systems.* Cambridge University Press, UK, 1998.
3. M. Gardner. The fantastic combinations of Jhon Conway's new solitaire game of life. *Scientific American*, 220, 1970.
4. S. Wolfram. *Theory and Application of Cellular Automata.* World Scientific, 1986.
5. S. Wolfram. *Cellular Automata and Complexity.* Reading, MA, 1994.
6. J. Hardy, O. de Pazzis, and Y. Pomeau. Molecular dynamics of a classical lattice gas: Transport properties and time correlation functions. *Phys. Rev. A*, 13:1949–1961, 1976.
7. U. Frisch, B. Hasslacher, and Y. Pomeau. Lattice-gas automata for the Navier-Stokes equation. *Physical Review Letters*, 56:1505–1508, 1986.
8. J. P. Rivet and J. P. Boon. *Lattice Gas Hydrodynamics.* Cambridge University Press, UK, 2001.
9. G. R. McNamara and G. Zanetti. Use of the Boltzmann equation to simulate lattice-gas automata. *Phys. Rev. Lett.*, 61:2332–2335, 1988.
10. S. Chen and G. D. Doolen. Lattice Boltzmann method for fluid flows. *Annual Review of Fluid Mechanics*, 30:329–364, 1998.
11. F. Higuera and J. Jiménez. Boltzmann approach to lattice gas simulations. *Europhys lett.*, 9:663–668, 1989.
12. Y. H. Qian. *Lattice Gas and lattice kinetic theory applied to the Navier-Stokes equations.* PhD thesis, Université Pierre et Marie Curie, Paris, 1990.
13. S. Chen, H. D. Chen, D. Martinez, and W. Matthaeus. Lattice Boltzmann model for simulation of magnetohydrodynamics. *Phys. Rev. Lett.*, 67:3776–3779, 1991.
14. P. L. Bhatnagar, E. P. Gross, and M. Krook. A model for collision processes in gases. i: small amplitude processes in charged and neutral one-component system. *Phys. Rev.*, 94:511–525, 1954.
15. J. G. Zhou. A lattice Boltzmann model for the shallow water equations. *Comp. Meth. Appl. Mech. Eng.*, 191(32):3527–3539, 2002.
16. J. G. Zhou. A lattice Boltzmann model for the shallow water equations with turbulence modelling. *International Journal of Modern Physics C*, 13:1135–1150, 2002.

17. J. G. Zhou. An elastic-collision scheme for lattice Boltzmann methods. *International Journal of Modern Physics C*, 12:387–401, 2001.
18. J. G. Zhou. A centred scheme for force terms in lattice Boltzmann methods. (Submitted), 2002.
19. J. H. Ferziger and M Perić. *Computational Methods for Fluid Dynamics*. Springer, Germany, 1996.
20. M. Tutar and A. E. Holdø. Computational modelling of flow around a circular cylinder in sub-critical flow regime with various turbulence models. *Int. J. Num. Meth. Fluids*, 35:763–784, 2001.
21. J. Smagorinsky. General circulation experiments with the primitive equations. *Monthly Weather Review*, 91:99–152, 1963.
22. P. K. Stansby and J. G. Zhou. Shallow-water flow solver with non-hydrostatic pressure: 2D vertical plane problems. *Int. J. Num. Meth. Fluids*, 28:541–563, 1998.
23. R. A. Falconer. An introduction to nearly-horizontal flows. In M. B. Abbott and W. A. Price, editors, *Coastal, Estuarial and Harbour Engineer's Reference Book*, pages 27–36, London, UK, 1993. Chapman and Hall.
24. I. S. Sokolnikoff and R. M. Redheffer. *Mathematics of Physics and Modern Engineering*. McGraw-Hill, New York, USA, 2nd edition, 1966.
25. M. Spivak. *Calculus*. W. A. Benjamin Inc., New York, USA, 1967.
26. J. Kuipers and C. B. Vreugdenhil. Calculations of two-dimensional horizontal flow. Tech. rep. s163 part 1, Delft Hydraulics Laboratory, Delft, The Netherlands, 1973. pp 1-44.
27. J. J. McGuirk and W. Rodi. A depth-averaged mathematical model for the field of the near side discharges into open-channel flow. *J. Fluid Mech.*, 86:761–781, 1978.
28. C. B. Vreugdenhil and J. H. Wijbenga. Computation of flow patterns in rivers. *J. Hydr. Eng. Div., ASCE*, 108(11):1296–1310, 1982.
29. A. G. L. Borthwick and G. A. Akponasa. Reservoir flow prediction by contravariant shallow water equations. *J. Hydr. Eng. Div., ASCE*, 123(5):432–439, 1997.
30. R. E. Featherstone and C. Nalluri. *Civil Engineering Hydraulics*. Hartnolls Ltd, Bodmin, Cornwall, Great Britain, 3rd edition, 1995.
31. J. G. Zhou. Velocity-depth coupling in shallow water flows. *J. Hydr. Eng., ASCE*, 121(10):717–724, 1995.
32. V. Casulli. Semi-implicit finite difference methods for the two-dimensional shallow water equations. *J. Comput. Phys.*, 86:56–74, 1990.
33. E. F. Toro. Riemann problems and the WAF method for solving two-dimensional shallow water equations. *Phil. Trans. Roy. Soc. Lond., A*, 338:43–68, 1992.
34. J. G. Zhou, D. M. Causon, C. G. Mingham, and D. M. Ingram. The surface gradient method for the treatment of source terms in the shallow-water equations. *J. Comput. Phys.*, 168:1–25, 2001.
35. R. Benzi, S. Succi, and M. Vergassola. The lattice Boltzmann equation: Theory and applications. *Physics Reports*, 222:45–197, 1992.
36. D. R. Noble, S. Chen, J. G. Georgiadis, and R. O. Buckius. A consistent hydrodynamic boundary condition for the lattice boltzmann method. *Physics of Fluids*, 7:203–209, 1995.
37. P. A. Skordos. Initial and boundary conditions for the lattice Boltzmann method. *Physical Review E*, 48:4823–4842, 1993.

38. X. He and L.-S. Luo. A priori derivation of lattice Boltzmann equation. *Physical Review E*, 55:R6333–R6336, 1997.
39. D. H. Rothman and S. Zaleski. *Lattice-Gas Cellular Automata*. Cambridge University Press, London, 1997.
40. H. Chen, S. Chen, and W. H. Matthaeus. Recovery of the Navier-Stokes equations using a lattice-gas Boltzmann model. *Physical Review A*, 45:R5339–R5342, 1992.
41. G. W. Yan. A lattice Boltzmann equation for waves. *J. Comput. Phys.*, 161:61–69, 2000.
42. J. D. Sterling and S. Chen. Stability analysis of lattice Boltzmann methods. *Journal of Computational Physics*, 123:196–206, 1996.
43. T. Abe. Derivation of the lattice Boltzmann method by means of the discrete ordinate method for the Boltzmann equation. *J. Comput. Phys.*, 131:241–246, 1997.
44. J. M. V. A. Koelman. A simple lattice Boltzmann scheme for Navier-Stokes fluid flow. *Europhys lett.*, 15:603–607, 1991.
45. D. Kandhai, W. Soll, S. Chen, A. Hoekstra, and P.Sloot. Finite difference lattice BGK methods on nested grids. *Computer Physics Communications*, 129:100–109, 2000.
46. D. Yu, R. Mei, and W. Shyy. A multi-block lattice Boltzmann method for viscous fluid flows. *Int. J. Num. Meth. Fluids*, 39:99–120, 2002.
47. C. L. Lin and Y. G. Lai. Lattice Boltzmann method on composite grids. *Phys. Rev. E*, 62:2219–2225, 2000.
48. N. S. Martys, X. Shan, and H. Chen. Evaluation of the external force term in the discrete Boltzmann equation. *Physical Review E*, 58(5):6855–6857, 1998.
49. J. M. Buick and C. A. Greated. Gravity in a lattice Boltzmann model. *Physical Review E*, 61(5):5307–5320, 2000.
50. S. Hou, J. Sterling, S. Chen, and G. D. Doolen. A lattice Boltzmann subgrid model for high Reynolds number flows. *Fields Institute Communications*, 6:151–166, 1996.
51. J. J. Quirk. An alternative to unstructured grids for computing gas-dynamic flows around arbitrarily complex 2-dimensional bodies. *Computers and Fluids*, 23(1):125–142, 1994.
52. J. G. Zhou and I. M. Goodwill. A finite volume method for steady state 2-d shallow water flows. *Int. J. Num. Meth. Heat Fluid Flow*, 7(1):4–23, 1997.
53. Q. Zou and X. He. On pressure and velocity boundary conditions for the lattice Boltzmann BGK model. *Physics of Fluids*, 9:1591–1598, 1997.
54. N. Goutal and F. Maurel, editors. *Proceedings of the 2nd Workshop on Dambreak Wave Simulation*, HE-43/97/016/B. Départament Laboratoire National d'Hydraulique, Groupe Hydraulique Fluviale, Electricité de France, France, 1997.
55. M. E. Vázquez-Cendón. Improved treatment of source terms in upwind schemes for shallow water equations in channels with irregular geometry. *J. Comput. Phys*, 148:497–526, 1999.
56. A. Bermudez and M. E. Vázquez. Upwind methods for hyperbolic conservation laws with source terms. *Computers and Fluids*, 23:1049–1071, 1994.
57. B. Yulistiyanto, Y. Zech, and W. H. Graf. Flow around a cylinder: shallow-water modeling with diffusion-dispersion. *J. Hydr. Eng., ASCE*, 124(4):419–429, 1998.

58. R. J. LeVeque. Balancing source terms and flux gradients in high-resolution Godunov methods: The quasi-steady wave-propagation algorithm. *J. Comput. Phys*, 146:346–365, 1998.
59. K. Shiono and D. W. Knight. Turbulent open-channel flows with variable depth across the channel. *J. Fluid Mech.*, 222:617–646, 1991.
60. P. M. Lloyd and P. K. Stansby. Shallow-water flow around model conical island of small side slope. II: submerged. *J. Hydr. Eng., ASCE*, 123(12):1068–1077, 1997.
61. B. Rogers, M. Fujihara, and A. G. L. Borthwick. Adaptive Q-tree Godunov-type scheme for shallow water equations. *Int. J. Num. Meth. Fluids*, 35:247–280, 2001.
62. C. Kranenburg. Wind-driven chaotic advection in a shallow model lake. *J. Hydr. Res.*, 30(1):29–46, 1992.
63. I. L. Rozovskii. *Flow of Water in Bends of Open Channels*. Israel Program for Scientific Translation, Jerusalem, Israel, 1965.

Index